GLOBAL ENVIRONMENTAL CHANGE

Understanding the Human Dimensions

Paul C. Stern, Oran R. Young, and Daniel Druckman, Editors

Committee on the Human Dimensions of Global Change
Commission on the Behavioral and Social Sciences
and Education
National Research Council

NATIONAL ACADEMY PRESS
Washington, D.C. 1992

NATIONAL ACADEMY PRESS • 2101 Constitution Avenue, N.W. • Washington, D.C. 20418

NOTICE: The project that is the subject of this report was approved by the Governing Board of the National Research Council, whose members are drawn from the councils of the National Academy of Sciences, the National Academy of Engineering, and the Institute of Medicine. The members of the committee responsible for the report were chosen for their special competences and with regard for appropriate balance.

This report has been reviewed by a group other than the authors according to procedures approved by a Report Review Committee consisting of members of the National Academy of Sciences, the National Academy of Engineering, and the Institute of Medicine.

Front Cover: Adapted from a photograph of the earth taken by Apollo 16 astronauts.

Library of Congress Cataloging-in-Publication Data

Global environmental change : understanding the human dimensions /
   Paul C. Stern, Oran R. Young, and Daniel Druckman, editors ;
   Committee on the Human Dimensions of Global Change, Commission on
   the Behavioral and Social Sciences and Education, National Research
   Council.
         p.   cm.
   Includes bibliographical references and index.
   ISBN 0-309-04494-4
   1. Man—Influence on nature. 2. Human ecology—Research.
3. Climatic change. 4. Environmental protection—Research.
5. Environmental policy. I. Stern, Paul C., 1944– . II. Young,
Oran R. III. Druckman, Daniel, 1939– . IV. National Research
Council (U.S.). Committee on the Human Dimensions of Global Change.
GF75.G57 1992
363.7—dc20   91-35182
                                                                  CIP

Printed in the United States of America

First Printing, December 1991
Second Printing, February 1993

iii

The National Academy of Sciences is a private, nonprofit, self-perpetuating society of distinguished scholars engaged in scientific and engineering research, dedicated to the furtherance of science and technology and to their use for the general welfare. Upon the authority of the charter granted to it by the Congress in 1863, the Academy has a mandate that requires it to advise the federal government on scientific and technical matters. Dr. Frank Press is president of the National Academy of Sciences.

The National Academy of Engineering was established in 1964, under the charter of the National Academy of Sciences, as a parallel organization of outstanding engineers. It is autonomous in its administration and in the selection of its members, sharing with the National Academy of Sciences the responsibility for advising the federal government. The National Academy of Engineering also sponsors engineering programs aimed at meeting national needs, encourages education and research, and recognizes the superior achievements of engineers. Dr. Robert M. White is president of the National Academy of Engineering.

The Institute of Medicine was established in 1970 by the National Academy of Sciences to secure the services of eminent members of appropriate professions in the examination of policy matters pertaining to the health of the public. The Institute acts under the responsibility given to the National Academy of Sciences by its congressional charter to be an adviser to the federal government and, upon its own initiative, to identify issues of medical care, research, and education. Dr. Stuart Bondurant is acting president of the Institute of Medicine.

The National Research Council was organized by the National Academy of Sciences in 1916 to associate the broad community of science and technology with the Academy's purposes of furthering knowledge and advising the federal government. Functioning in accordance with general policies determined by the Academy, the Council has become the principal operating agency of both the National Academy of Sciences and the National Academy of Engineering in providing services to the government, the public, and the scientific and engineering communities. The Council is administered jointly by both Academies and the Institute of Medicine. Dr. Frank Press and Dr. Robert M. White are chairman and vice chairman, respectively, of the National Research Council.

Preface

 B ecause human activities interact with physical and biological systems both as driving forces and as critical links in feedback mechanisms, any effort to understand, much less to come to terms with, global environmental change that does not include a sustained commitment to improving our knowledge of the human dimensions cannot succeed. Awareness of this simple truth is now spreading throughout the scientific community. In its report prepared to accompany the fiscal 1992 federal budget request for global change research, to take a prominent example, the Committee on Earth and Environmental Sciences flatly states that "[w]ithout an understanding of human interactions in global environmental change that is based both on empirical observations of human behavior and on a better understanding of the consequences of human actions, the models of physical and biological processes of change will be incomplete." As the committee goes on to observe, an especially critical need is the "identification of the ways that human, physical, and biological systems interact, often through complex feedback mechanisms."

What are the implications of this realization for research on the human dimensions of global environmental change? Can we expect to understand the proximate causes of anthropogenic change— releases of carbon dioxide and chlorofluorocarbons (CFCs), transformation of forests into fields, and so forth—without digging deeper to analyze the underlying sources of the human behavior in question? How should we allocate our attention between studies of the underlying sources of anthropogenic change and studies of human responses to global environmental change, which figure prominently in the relevant feedback mechanisms? How can we exploit existing social data in order to advance our understanding of the human dimensions of global change, and how can we set priorities to make the best use of any new resources that become available to collect and perfect data in this field of inquiry? What can we do to break down intellectual and institutional barriers between the social sciences and the natural sciences or, for that matter, among the individual social science disciplines in the interests of deepening our understanding of global environmental change?

In 1989, the National Research Council (NRC), with support from the National Science Foundation, the John D. and Catherine T. MacArthur Foundation, the National Research Council Fund, and the U.S. Geological Survey, established the Committee on the Human Dimensions of Global Change as a means of responding to these questions. The committee's charge included four interrelated tasks. Specifically, the committee was asked to undertake: "an assessment of previous social science research on topics related to global change; an evaluation of extant data resources for social and behavioral research on global change; a consideration of how collaborative research on global change might influence the generation of knowledge in the social sciences as well as attract social and behavioral scientists to apply their knowledge to global issues; and the development of a research agenda that can be implemented over a period of several years."

The Committee on the Human Dimensions of Global Change has not labored alone in endeavoring to identify and to grasp the human component of the complex interactions among human, physical, and biological systems giving rise to global change. The Committee on Global Change (the organizing committee for U.S. participation in the International Geosphere-Biosphere Programme) has formed a Human Interactions Panel; the Social Science Research Council has set up a Committee for Research on Global Environmental Change; and the National Science Foundation has initiated a program of investigator-initiated grants on the Human Dimensions of Global Environmental Change. Nor is this activity confined to the United States. Most notably, the International Social Science Council has put in place a Standing Committee on the Human Dimensions of Global Change; this committee is now moving vigorously to make common cause with the International

Geosphere-Biosphere Programme in the interests of stimulating collaboration between natural scientists and social scientists working on global change issues.

What sets the work of the Committee on the Human Dimensions of Global Change apart from these other endeavors, it seems to me, is the sustained effort the committee has made to build a solid foundation under human dimensions research as a coherent intellectual enterprise and a legitimate field of study that can command the resources needed to advance understanding in important ways. This means, to begin with, laying out systematically the parameters, premises, and problems of this field of study rather than restricting attention to the identification of a loose collection of substantive topics deemed worthy of attention on the part of those interested in global change. It means, as well, addressing a series of issues involving human resources and organization matters that cannot be ignored if this field of study is to establish itself as a going concern within the scientific community. The report that follows contains numerous recommendations. In the end, it lays out a five-point plan for a national research program on the human dimensions of global environmental change. But the persuasiveness of this plan rests, in the final analysis, on the success of our effort to build a foundation for human dimensions of global environmental change as a coherent field of study.

Guiding this project has been a challenging and stimulating endeavor from the moment the process of selecting committee members began to the day the last suggestions of a battery of reviewers were considered. Because it is hard to think of anything that has no bearing on global environmental change, we assembled a highly diverse group of senior scientists to conduct the work of the committee. Through most of its deliberations, the committee has consisted of 15 members, 12 representing a wide range of social science disciplines (including history and law) and 2 drawn from the natural sciences (including the life sciences and physical sciences). The result, predictably enough, has been a series of spirited exchanges featuring elements of disciplinary firmness as well as mutual education. Our product reflects a sustained commitment on the part of all participants to the creation of a new field of study drawing on the intellectual capital of numerous disciplines but dominated by none.

In my experience, the effort required to bring this project to a successful conclusion has been extraordinary. It is no exaggeration, therefore, to say that we could not have produced this report without a commitment going beyond the call of duty on the part of Dan Druckman, Paul Stern, and Michelle Daniels, the staff members of the NRC's Commission on Behavioral and Social Sciences and Education assigned to this project. Working under considerable time pressure, Michelle made all our meetings run like clockwork. Paul was indefatigable both as the principal author of large sections of the report and as the staff member responsible both for integrating suggestions of outside reviewers and for handling a number of the tedious tasks involved in transforming a draft into a finished report. Dan was responsible for preparing the proposal for the project and making arrangements for its implementation. He was also a dependable source of good judgment on matters of substance, editing, and personnel as well as an active participant in our sessions. To all three, I offer both my sincere thanks for their unfailing support and my commendation for a job well done. In addition, special gratitude is extended to Christine McShane, the commission's editor: her skillful editing of the entire manuscript contributed substantially to its readability. Thanks go also to Donna Reifsnider for her administrative support. Definition and guidance for the committee's task came primarily from Roberta Miller, director the National Science Foundation's Division of Economic and Social Science. She was an important source of encouragement for the committee's work at all stages of the project.

Oran R. Young
Chair, Committee on the Human
Dimensions of Global Change

Contents

GLOBAL ENVIRONMENTAL CHANGE

Summary

The earth has entered a period of hydrological, climatological, and biological change that differs from previous episodes of global change in the extent to which it is human in origin. To explain or predict the course of the present global environmental changes, one must therefore understand the human sources, consequences, and responses, some of which can alter the course of global change. This book examines what is known about the human dimensions of global environmental change, identifies the major immediate needs for knowledge, and recommends a strategy for building that knowledge over the next 5-10 years.

To understand global environmental change, it is necessary to focus on the interactions of environmental systems, including the atmosphere, the biosphere, the geosphere, and the hydrosphere, and human systems, including economic, political, cultural, and sociotechnical systems. Human systems and environmental systems meet in two places: where human actions proximately cause environmental change, that is, where they directly alter aspects of the environment, and where environmental changes directly affect what humans value. The main questions about human causes concern the underlying sources or social driving forces that give rise to the proximate causes of global change. Why, for example, is there so much variation across societies, even advanced industrial societies, with regard to energy consumption per unit of economic output? The key questions about human consequences concern responses to actual or anticipated global changes. What will humans do in anticipation of global change to keep it from harming what they value? How will humans respond to actual global changes? What is the likelihood that humans will take no organized action at all in response to particular global changes, and what would be the consequent effects on human welfare? To answer such questions, natural and social scientists need to work together.

HUMAN CAUSES OF GLOBAL CHANGE

Almost all human activity has some potential relevance to global change. Researchers in a number of fields have studied human–environment interactions, usually within the boundaries of single disciplines and almost always below the global level. They have demonstrated that a complex of social, political, economic, technological, and cultural variables, sometimes referred to as driving forces, influences the human activities that proximately cause global change. The driving forces can be roughly classified as follows:

- *Population Growth* Each person makes some demand on the environment for the essentials of life—food, water, clothing, shelter, and so on. If all else is equal, the greater the number of people, the greater the demands placed on the environment for the provision of resources and the absorption of waste and pollutants. However, all else is not equal. For example, a new individual with the standard of living and technological base of an average North American would use about 35 times as much energy as an individual living at India's average standard—with a roughly proportional impact on the global environment.
- *Economic Growth* For the first time in human history, economic activity is so extensive that it produces environmental change at the global level; the prospect of further economic growth arouses concern about the quality of the global environment. Economic growth necessarily stresses the environment, but the amount of stress from a given amount of economic growth depends, among other things, on the pattern of goods and services produced, the population and resource base for agricultural development, forms of national political organization, and development policies.
- *Technological Change* Technology can influence environmental change by finding new ways to discover and exploit natural resources or by changing the volume of resources required—or the

amount or kind of wastes produced—per unit of output. Technologies may either increase or decrease the impact of human activity on the environment, depending on the other driving forces, which determine which technologies are developed and used.

• *Political-Economic Institutions* The global environment responds to the actions of markets, governments, and the international political economy. Markets are always imperfect, and the impact of economic activity on the environment depends on which imperfect-market method of environmental management is being used. Governmental structure and policies can also have significant environmental consequences, both intentional and inadvertent. And the international political economy, with its global division of labor and wealth, can promote environmental abuses, particularly in the Third World. The effects depend on policy at the national level and on the behavior of particular economic actors.

• *Attitudes and Beliefs* Beliefs, attitudes, and values related to material possessions and the relation of humanity and nature are often seen as lying at the root of environmental degradation. Such attitudes and beliefs probably have their greatest independent effects over the long term, on the time scale of human generations or more. Within single lifetimes, attitudes and beliefs can have significant influence on resource-using behavior, even when social-structural and economic variables are held constant.

Although each of these driving forces is important at certain times and under certain conditions, much remains unknown about what determines their relative importance, how they affect each other, and how the driving forces in particular places combine to produce global effects. For example, various combinations of social conditions may lead to a single outcome, such as deforestation.

Single-factor explanations of the anthropogenic sources of global environmental change are apt to be misleading, because the driving forces of global change generally act in combination with each other and the interactions are contingent on place, time, and level of analysis. Understanding the linkages is a major scientific challenge that will require developing new interdisciplinary teams. The research effort should include studies at both global and lower geographic levels, with strong emphasis on comparative studies at local or regional levels with worldwide representation. Research should address the same question at different time scales, examine the links between levels of analysis and between time scales, and explore the ways that the human forces that cause environmental change may also be affected by it.

HUMAN CONSEQUENCES OF GLOBAL CHANGE

To project the human consequences of global change, it is necessary not only to anticipate environmental change but also to take social change into account: social and economic organization and human values may change faster than the global environment, and people may respond in anticipation of global change. It is worthwhile to test projected environmental futures against projected human futures to see how sensitive human consequences may be to variations in the social future. But long-term forecasting is still a very inexact practice. The near-term research agenda should emphasize processes of human response to the stresses that global environmental change might present.

People may respond to experienced or anticipated global change by intervening at any point in the cycle of interaction between human and environmental systems. Mitigation—that is, actions that alter environmental systems to prevent, limit, delay, or slow the rate of undesired global changes—may involve direct interventions in the environment, direct interventions in the human proximate causes, or indirect interventions in the driving forces of global change. People can also respond by blocking the undesired proximate effects of environmental systems on what they value, for example, by applying sunscreens to the skin to help prevent cancer from exposure to ultraviolet radiation. They can make adjustments that prevent or compensate for imminent or manifest losses of welfare from global change, for example, famine relief or drought insurance. And people can intervene to improve the robustness of social systems by altering them so that an unchecked environmental change would produce less reduction of values than would otherwise be the case. For example, crop polyculture may not slow the pace of global change, but it is more robust than monoculture in the face of drought, acid deposition, and ozone depletion. If crop failure occurs, it will affect only some crops, making famine less likely. Many of these responses may indirectly affect the driving forces of global change. Consequently, the research agenda should include studies of both the direct and secondary effects of responses to global change, using the best available methods of evaluation research.

Global change is likely to engender conflict—about whether it is in fact occurring, whether any organized response is necessary, whether response should emphasize mitigation or not, who should pay the costs, and who has the right to decide. Such conflicts tend to persist because they are based in part on differing interests, values, preferences, and beliefs about the future. The research agenda should include efforts to clarify the connections between particular environmental changes and particular types of conflict. It should also include increased efforts to test the efficacy of different techniques and institutions for resolving or managing environmental conflicts.

Human responses to global change occur within **seven** interacting systems. Within each system, there are significant areas of knowledge and important unanswered questions; in addition, much remains to be learned about how the systems combine to determine the global human response.

• *Individual perception, judgment, and action* are important because all decisions are based on inputs from individuals; because individual actions, in the aggregate, often have major effects; and because individuals can be organized to influence collective and political responses.

• *Markets* are important because global change is likely to affect the prices of important commodities and factors of economic production. However, existing markets do not provide the right price signals for managing global change for various reasons, and the participants in markets do not always follow strict rules of economic rationality.

• *Sociocultural systems*, including families, clans, tribes, and communities held together by such bonds as solidarity, obligation, duty, and love sometimes develop ways of interacting with their environments (for instance, some systems of agroforestry) that may be widely adaptable as strategies for response. Their informal social bonds can also affect individual and community responses to global change and to policy.

• *Organized responses at the subnational level*, such as by communities, social movements, and corporations and trade associations, can be significant both in their own right and by influencing the adoption and implementation of government policies.

• *National policies* are critical in the human response to global change by making possible international agreements and by affecting the ability to respond at local and individual levels. Not only environmental policy, but also macroeconomic, fiscal, agricultural, and science and technology policies are important.

• *International cooperation* is necessary to address some large-scale environmental changes such as ozone depletion and global warming. The formation of international institutions for response to global change is widely considered to be the key to solving the problems, and both nation-states and non-state actors are involved.

• *Global social change*, such as the expansion of the global market; the worldwide spread of communication networks, democratic political forms, and scientific knowledge; and the global resurgence of cultural identity as a social force may influence the way humanity responds to the prospect of global change and its ability to adapt to experienced change.

The research agenda should include studies of responses within each of these systems, especially comparative studies of how the systems operate in different spatial and temporal contexts. Because systems of human response are strongly affected by each other, a high priority should be given to studies linking response systems to each other and short-term effects to long-term ones.

PROBLEMS OF THEORY AND METHOD

The study of human interactions with the global environment poses difficult problems of theory and method that will require new links among disciplines, theoretical constructs to deal with the complexities and the large spatial and temporal scales, and careful selection of research methods.

• *Interdisciplinary collaboration* is essential. A high priority of the human interactions research effort should be to support problem-centered interaction among social and natural scientists, for example through research projects that require such contact, problem-focused scholarly meetings, and interdisciplinary research centers.

• *New theoretical tools* are required. Studies of the human dimensions of global change require analysis at spatial and temporal expanses much greater than most social scientific theory encompasses. Social science will need to develop new theoretical tools for analyzing such issues as major national and international changes in political-economic structure, the sources of variation and change in slowly changing aspects of human systems, the long-term impacts of short-term social changes, relationships between global social changes and the global environment, and links between human–environment relationships at different levels of spatial aggregation. Analyses of these general problems in the global change context may lead to important theoretical advances of general use in social science.

• *Methodological pluralism* is the most appropriate strategy. At least for the near term, a strong emphasis on building integrative models is premature for studying the human dimensions. For the human interactions research agenda, much more understanding of the underlying processes needs to be developed before great strides can be made in integrative modeling. Formal modeling should participate in a dialogue of methods, with several complementary methods being used to give a more complete picture than any single method can produce.

• *Post hoc analyses* are essential for evaluating human responses. There remains no substitute for empirical analysis of outcomes after the fact. Post hoc evaluations are an important part of the process of analyzing policy alternatives for response to global change, and resources should be provided for them. In particular, federal agencies with programs or policies anticipated to affect processes of global environmental change should routinely budget funds for evaluation studies of the intended and unintended consequences of these activities. Unlike the practice of preparing environmental impact statements, this recommendation concerns data gathering after a policy is in place.

DATA NEEDS

A strong research program on the human dimensions of global change requires improved availability of and access to existing data, quality control, and collection of critical new data.

• *Data Availability* Data exist in great quantity on social, economic, demographic, and political variables relevant to the human dimensions of global change, and in even greater quantity on relevant nonhuman variables. The major need at this point is for governmental and private support for the necessary infrastructure for publicly shared data on demographic, economic, political, attitudinal, and natural or physical changes. This particularly includes the one-time costs of creating a network and archival facility to make the material accessible to researchers and to make data on social and natural phenomena mutually intelligible. The network should include measures of the major driving forces of global change at the lowest available level of aggregation. We emphasize that expenditures on such a system should not jeopardize needed resources for doing research and understanding the data. The U.S. government should take steps to keep the price of basic data close to the marginal cost of production and should try to influence international institutions such as the Organisation for Economic Co-operation and Development (OECD) to do the same. The federal government should support an effort to validate the most promising remotely sensed indicators of social phenomena and include them in the information network. The committee recommends that social scientists, representing a variety of disciplines, be involved at every stage of the design and implementation of national data and information systems relevant to the human dimensions of global change, including representation on the Earth Observing System Science Advisory Panel, to ensure that data are collected and archived in a form that facilitates analyzing human interactions.

• *Quality and Interpretability of Data* The quality of existing data relevant to the human dimensions of global change may be doubtful because of errors in collection, problems of sampling and coverage, problems of estimation, incompatibility between ground-based and remotely sensed data, problems of aggregation, insufficient attention to methodology, or a lack of uniform definitions of variables across data-collection agents. The prevalence of these problems suggests the importance of research on the quality of available data sets. A few targeted studies that trace the production of data on key variables using modern techniques to analyze multiple indicators would greatly enhance understanding of data quality and would suggest methods for improving both conceptualization and measurement. The latter methods can usefully be applied to unreliable data on the physical and biological, as well as social, aspects of global change.

• *Needs for New Data* Needs for new data reveal themselves as research proceeds. Nevertheless, an inventory of existing data should be developed to determine whether expanded data collection is needed in such areas as land use and food production, economic activity, consumption of energy and materials, human health, population trends, environmental quality, and environmental attitudes. There may be needs for missing data at local, regional, or national levels; for improved aggregated data, such as on national income in developing countries and current or formerly socialist countries; for data on variables for which adequate measures do not yet exist; and for data on particular areas that are selected for purposes of comparison. The data inventory should be developed through consultation among researchers, governmental and nongovernmental statistical agencies, and data base management and archiving specialists; it should be reviewed and updated periodically.

HUMAN RESOURCES AND
ORGANIZATIONAL REQUIREMENTS

To develop an effective research program on the human dimensions of global change, critical needs must be met for improved institutional infrastructure, training and retraining, and a restructuring of the federal research effort to overcome barriers to environmental social science.

• *Institutional Structures* Global change research, especially in universities, faces the same serious barriers as other interdisciplinary programs: limited institutional support; small budgets; and few if any faculty appointments, particularly with tenure. Individuals who commit time to such programs often do so to the detriment of their own careers. Therefore, in addition to short-term research support, programs of research on the human dimensions of global change need to develop long-term institutional identities. The committee recommends that sponsors of research on the human dimensions of global change address some of their support to building institutional entities that control their own faculty appointments and other key resources that will enable them to attract the interest and resources of individuals who are already present but not yet committed to global change as a research agenda.

One way to address the problems of establishing a new area of interdisciplinary research is to create national centers for research. Centers on the human dimensions of global change could be funded by a consortium of government and private sources that would make a commitment to maintain them on a long-term basis. They should be rooted in environmental social science but should also maintain a commitment to collaborative work with natural scientists.

Incentives are also needed to encourage collaboration between natural and social scientists. Any research proposal that includes only natural scientists or only social scientists should be required to justify its decision, and grant review panels should be designed to ensure that disciplinary criteria do not bias evaluations of interdisciplinary proposals.

• *Training* Funders should make special efforts to promote interdisciplinary communication and cooperation with fellowship and travel grants and possibly by holding annual meetings of graduate students and postdoctoral fellows from different disciplines to build a sense of community and collegiality. Training efforts should also consider the fact that for most young researchers, career development will require that their interdisciplinary work be grounded in a disciplinary framework that potential employers will recognize. We recommend that proposals for substantive research, especially for graduate and postdoctoral research, be evaluated on their ability to synthesize interdisciplinary questions about global change with the theoretical and programmatic agendas of existing disciplines. The committee also recommends that federal and private funding sources make resources available on a competitive basis to professional associations to initiate programs to strengthen the links between the core theoretical concerns of individual disciplines and the human dimensions of global change.

• *Organizational Barriers to Research in the Federal Government* Due to the historical missions of federal agencies, there is an almost complete mismatch between the roster of agencies that support research on global change and the roster of agencies with strong capabilities in social science. Consequently, with the exception of the National Science Foundation, no entity in the federal government has the expertise to develop and manage a comprehensive research agenda on human interactions with the global environment, and many important research needs are likely to go unmet for lack of an agency with the mission and personnel needed to meet them. The federal government should develop a strategy to ensure that the human dimensions research agenda is designed and administered by organizations committed to excellence in understanding both environmental and human systems. The Committee on Earth and Environmental Sciences might, as appropriate, assign important areas of human interactions research to the National Science Foundation, to particular mission agencies with the requirement that they take on new staff or make use of outside expertise to handle the assignment, or, if no existing agency is appropriate, to a new organizational entity, staffed with social and natural scientists.

RECOMMENDATIONS FOR A
NATIONAL RESEARCH PROGRAM

The social and behavioral sciences have a vital contribution to make to enhancing understanding of global environmental change. This contribution can best be made through an effective partnership between the natural sciences and the social and behavioral sciences, but two key problems obstruct development of a strong research program: intellectual and organizational barriers to interdisciplinary collaboration and inappropriate organization within the U.S. government for managing the research effectively. The committee recommends that the United States develop a comprehensive national research program on the human dimensions of global change consisting of five major elements: investigator-initiated research, targeted or focused research on selected topics, a federal program for obtaining and disseminating relevant data, a program of fellowships to expand the pool of talented scientists in the field, and a network of national research centers.

Recommendation 1 **The National Science Foundation should increase substantially its support for investigator-initiated or unsolicited research on the human dimensions of global change. This program should include a category of small grants subject to a simplified review procedure.**

The National Science Foundation program of investigator-initiated research on the human dimensions of global change should be established on a long-term basis, structured to include the full range of social and behavioral sciences, and expanded substantially in terms of funding. The following evaluation criteria should be applied in selecting among high-quality proposals and should inform the thinking of those preparing proposals for submission.

a. Studies of the anthropogenic sources of global change deserve priority to the extent that they address human actions that have a large impact on one or more of the major global environmental changes.

b. Studies of the anthropogenic sources of global change should receive priority to the extent that they emphasize interactions among social driving forces.

c. While there is a place for global-level studies, the emphasis in the near term should fall on comparative studies at the national, regional, and local levels. This approach will promote fuller understanding of the processes at work in human interactions with global change—an understanding that must form the basis for generalizations at the global level.

d. Although there is room for analyses on different time scales, there is a need to be especially supportive of studies dealing with the environmental effects of human actions on time scales of decades to centuries. Understanding global change requires an examination of long-term changes in human systems as well as environmental systems.

e. There is a need to support studies that compare interventions at different points in the cycle of human–environment relationships and make empirical assessments of their relative effects.

f. Research should make a systematic effort to compare and contrast the responses of human systems at different levels of social organization.

g. There is much to be gained from studies that differentiate among distinct methods or mechanisms for influencing human behavior.

h. There is a need for studies of the robustness of human systems (including social, technical, agricultural, economic, and political systems) in the face of global environmental change.

i. Proposals deserve priority to the extent that they are likely to enhance understanding of processes of decision making and conflict management in response to global environmental changes. Given the widespread frustration associated with policy making concerning environmental issues and the magnitude of the human responses that may be needed, a concerted effort to improve the quality of collective decision making in this area is warranted.

j. Special attention should be given to proposals that suggest effective methods of enhancing the partnership between the natural sciences and the social sciences or encouraging interdisciplinary research among the social sciences relating to global environmental change.

k. Proposals deserve serious consideration to the extent that they include effective plans for increasing international collaboration.

> **Recommendation 2   The National Science Foundation, other appropriate federal agencies, and private funding sources should establish programs of targeted or focused research on the human dimensions of global change.**

There is a national need to establish an ongoing program of targeted or focused research—that is, a program that will concentrate resources to advance understanding of topics selected by the funding sources for their obvious significance for global environmental change. All topics selected for focused research should meet the following criteria: (i) they deal with matters of first-order significance to understanding causes, consequences, and responses to global environmental change, (ii) they raise questions that typify larger classes of concerns relating to the human dimensions of global change, (iii) they address the major categories of global environmental change (for example, ozone depletion, climate change, and the loss of biodiversity), and (iv) they show promise of yielding timely advances regarding questions of broad interest to the social sciences. The topics must also be sufficiently well defined to provide a basis for targeted research. Mission agencies that usually support only applied research in social science but have basic research programs in natural science related to global change, should initiate support of basic research in the social sciences related to their global change missions.

Following are examples of topics that meet our criteria for inclusion in the initial phase of focused programs dealing with the human dimensions of global change:

- *Energy Intensity* Why do economies differ so markedly in their energy intensity? How and why does the consumption of energy per unit of GNP change over time? What do the answers to these questions imply about opportunities to reduce carbon dioxide emissions?
- *Land Use and Food Production* What factors change systems of land use and food production toward either rapid degradation of resources or sustainability? How do such changes correlate with population growth, technological development, and the evolution of social institutions?
- *Valuing Consequences of Environmental Change* What alternative approaches can be used to place values on those consequences of environmental changes that are not well reflected in market prices? What institutional arrangements could ensure the effective use of the most promising of these approaches?
- *Technology–Environment Relationship* What determines whether the technologies developed and adopted in major economic sectors mitigate or exacerbate global environmental change? What are the roles of factor prices, regulatory practices, legal and institutional arrangements, standards of performance or practice, and other characteristics of the decision environment in determining which technological options are pursued and adopted?
- *Decision Making in Response to Global Environmental Change* How do individuals, firms, communities, and governments come to perceive changes in environmental systems as requiring action? How do they identify possible responses and assess the probable consequences of such responses? Are there cultural differences in the way human communities deal with such issues?
- *Environmental Conflict* How might global environmental

changes intensify existing social conflicts or engender new forms of conflict? What techniques of conflict resolution or conflict management are likely to prove effective in coming to terms with these conflicts?

• *International Environmental Cooperation*   What can we learn from the recent experience with ozone depletion that is relevant to international efforts to deal with climate change or the loss of biodiversity? When do governments resort to international cooperation in dealing with environmental changes, and when are the resultant regimes likely to prove effective?

> **Recommendation 3   The federal government should establish an ongoing program to ensure that appropriate data sets for research on the human dimensions of global change are routinely acquired, properly prepared for use, and made available to researchers on simple and affordable terms.**

There is a national need to (i) inventory existing data sets relevant to the human dimensions of global change, (ii) critically assess the quality of the most important of these data sets, (iii) make determinations concerning the quality of data required for research on major themes, (iv) investigate the cost-effectiveness of various methods of improving the quality of critical data sets, and (v) make decisions regarding new data needed to underpin a successful program of research. A federal program is warranted because public agencies collect the bulk of the relevant data and because the task is so large that individuals or private groups cannot hope to handle it effectively or efficiently. The federal government should seriously consider the establishment of a national data center on the human dimensions of global change parallel to the centers that already exist for data on climate, oceans, geophysics, and space science. An independent advisory committee, composed of researchers working on the human dimensions of global change and including strong representation of social scientists, should be set up to oversee the work of the federal data program.

> **Recommendation 4   The federal government, together with private funding sources, should establish a national fellowship program. Through it, social and natural scientists prepared to make a long-term commitment to the study of the human dimensions of global environmental change could spend up to two years interacting intensively with scientists from other disciplines, especially scientists from across the social science–natural science divide.**

It is imperative to find ways to allow individual scientists to push beyond the boundaries of their home disciplines in thinking about global change without jeopardizing their career trajectories. A prestigious nationwide fellowship can induce students and researchers to enhance their knowledge of global change issues and to interact intensively across disciplines. The fellowships should be open to graduate students, postdoctoral scientists, and mid-career scientists on a competitive basis and carry competitive stipends.

Recommendation 5 The federal government should join with private funding sources to establish about five national centers for research on the human dimensions of global change and to make a commitment to funding these centers on a long-term basis.

Because of the interdisciplinary, problem-oriented nature of the topic, the human dimensions of global change constitutes an emerging field of inquiry that is ripe for this sort of treatment; about five centers should be established over the next 3-5 years. National research centers should be established at locations that employ scientists with a proven track record in this or related areas, to avoid the problem of intellectual opportunism. In the committee's judgment, there is a persuasive case for maintaining relatively close ties between universities and the national centers. However, topics that lie at the intersection between basic and applied research may be most appropriately investigated at a government-operated research center.

Recommendation 6 The federal government should increase funding for research on the human dimensions of global change over a period of several years to a level of $45-50 million.

This cost estimate assumes that the program would be phased in over time and that, because all five program elements are necessary to the comprehensive national research program, the program will strike a proper balance among the program elements with regard to funding. We believe that investigator-initiated research on human interactions, currently funded at a level of $3.6 million per year through the National Science Foundation, can and should be tripled to a level of about $11 million. Targeted or focused research on the human dimensions of global change should be funded at a comparable level with investigator-initiated re search. A fellowship program in full operation that awarded 100 two-year fellowships per year would cost $10 million per year if the average annual cost were $50,000 per fellowship, including indirect costs. Five national centers could be maintained with small but strong core staffs for about $1 million per center per year. Funding for data acquisition and dissemination should remain, as in the fiscal 1991 budget, at about 20 percent of the funds allocated to human interactions research. On that basis, we recommend that funding for the data program be increased over the transition period to a level of $8-10 million.

The committee believes that this level of support would make possible the establishment of a balanced national research program on the human dimensions of global change and that the research community will be able to take on such a commitment over a three-year period if the funding is available. This level of funding would represent about 5 percent of the fiscal 1991 budget for the U.S. Global Change Research Program (USGCRP) or 4 percent of the proposed fiscal 1992 budget, in contrast to the 3 percent currently budgeted. In light of the National Research Council's conclusion that the human interactions science priority is "the most critically underfunded in the FY 1991 budget for the USGCRP" (National Research Council, 1990b:95), an increase of this magnitude over a short time period seems fully justified. Support for appropriate parts of the research program outlined here could come from an emerging Mitigation and Adaptation Research Strategies program as well as from the Global Change Research Program.

1
Prologue

The earth has entered a period of hydrological, climatological, and biological change that differs from previous episodes of global change in the extent to which it is human in origin. Human beings, both individually and collectively, have always sought to transform their surroundings. But for the first time, they have begun to play a central role in altering global biogeochemical systems and the earth as a whole. The global changes looming largest on the horizon are cases in point. The depletion of the ozone layer attributed to the accumulation of chlorofluorocarbons (CFCs) in the stratosphere is an unintended side effect of human industrial activities. The increase of atmospheric carbon dioxide, a trend that has been accelerating since the onset of the Industrial Revolution, is driven by the increasing use of fossil fuels and the elimination of forests. And the loss of biological diversity is a by-product of varied human activities, including the clearing of tropical moist forests for agricultural purposes.

To explain or predict the course of such global environmental changes, one must therefore understand the human sources of these changes. Because people engage in purposeful behavior, it is also important to focus on human responses to global changes. For example, after people learned that CFCs rising into the stratosphere would deplete the ozone there and threaten human health, they made serious efforts to phase out industrial and commercial uses of these chemicals. Such human responses can alter the course of global environmental change.

THREE GLOBAL CHANGES

The human behavior in question is both complex and poorly understood. To illustrate this point and to provide a backdrop for the analysis to come, we begin with three vignettes that convey a sense of the range of concerns addressed in this book. These three examples reappear later on to illustrate issues in the human dimensions of global change.

GREENHOUSE GASES AND CLIMATE CHANGE

Human activities threaten to alter the global climate by releasing so-called greenhouse gases, principally carbon dioxide, methane, CFCs, and nitrous oxide, that have the effect of increasing the proportion of heat from the sun that is retained at the top of the atmosphere. While it is easy to establish a connection between the onset of the Industrial Revolution and increases of carbon dioxide in the earth's atmosphere, the story of greenhouse gases is far more complicated. An examination of emissions of carbon dioxide, a by-product of the combustion of fossil fuels to produce energy and the most important of the anthropogenic greenhouse gases, suggests a clear link between economic development and a global increase in greenhouse gases. Yet there are striking disparities between otherwise similar economies in how much energy they use to produce a unit of economic output. Japan uses less than half the energy the United States does; Bangladesh uses about half what India or Pakistan does; the range in sub-Saharan Africa includes the least energy-intensive economy in the world (Lesotho) and one of the most intensive (Zambia), which differ by a factor of 56 (World Bank, 1989). Other human activities, such as the destruction of tropical moist forests, are major contributors to carbon dioxide emissions worldwide. In the case of methane, population pressure appears to be a key factor, since the expansion of rice paddies to feed the growing human populations of East and South Asia is a major source of methane emissions. The case of CFCs points to the role of technology as a source of the greenhouse effect; CFCs were developed initially to eliminate problems with existing refrigerants, like ammonia, rather than as a response to population pressure or the forces of economic development (although, of course, population growth and economic production add to the demand for CFCs). Nitrous oxide tells yet another tale. The largest single source of this greenhouse gas is the use of nitrogenous fertilizers to increase crop yields, mainly in advanced agricultural systems.

Chlorofluorocarbons and the Stratospheric Ozone Layer

In 1987 an international protocol was signed at Montreal, in which signatory countries declared their intention to cut CFC production and consumption in half by the end of the century. The Montreal Protocol is remarkable for the fact that it constitutes an anticipatory response to global change. Although held up by many as a model for those working on other global environmental issues, in its initial form this international agreement left much to be desired: it did not cover a range of chemicals (for example, methyl chloroform, carbon tetrachloride) whose effects are similar to those of CFCs; it allowed developing countries to actually increase their use of CFCs; it offered minimal assistance to governments seeking to reduce the use of CFCs within their own jurisdictions; it provided little guidance on compliance; and it did not succeed in drawing in key players like China and India. But the pressure of worldwide public opinion, driven by dramatic recurrences of sharp yearly reductions in ozone over Antarctica and by a growing scientific consensus concerning the dangers of ozone depletion, led to a renegotiation of the protocol that has strengthened its provisions in a number of areas. Under a series of amendments to the Montreal Protocol negotiated in London in 1990, CFCs and halons are to be phased out by the year 2000, and methyl chloroform and carbon tetrachloride have been added to the list of chemicals to be eliminated. The creation of an international fund to assist developing countries in switching to alternatives to CFCs has persuaded China and India to say they will join the agreement.

Nevertheless, the new provisions do not solve all the problems associated with ozone depletion. CFCs already in the atmosphere are expected to cause significant reductions in stratospheric ozone over a period of several decades. There is no guarantee that the chemicals developed as substitutes for CFCs will prove benign over the long run (CFCs themselves were thought to have ideal properties when they were introduced in the 1930s). Hydrochlorofluorocarbons (HCFCs) and hydrohalocarbons (HFCs), prominent among the candidates to be used as substitutes for CFCs, have already provoked opposition from those concerned about their potential contribution to the greenhouse effect. That opposition makes prospective producers such as DuPont reluctant to invest heavily in facilities needed to initiate large-scale production of the chemicals (Holusha, 1990a, b). Still, the international response to the problem of ozone depletion has been noteworthy as a quick and decisive reaction to a threat whose impact lies largely in the future. Many now wonder whether this case offers lessons of value for those struggling to come to terms with other global changes, such as climate change and the loss of biological diversity.

AMAZONIAN DEFORESTATION AND THE LOSS OF BIOLOGICAL DIVERSITY

Human activities are decreasing biological diversity on land, in fresh waters and the seas, in industrialized and developing nations, from the coldest inhabited lands to the tropics. But because of the huge variety of species confined to the tropical moist forests, their destruction is likely to cause more loss of biological diversity at the species level than any other human activity. Although the scale of this destruction is hard to measure accurately, recent estimates indicate that, on a world scale, an area of tropical moist forest roughly the size of Honduras is deforested or converted annually (Erwin, 1988), resulting in the extinction of species at a rate that has been estimated at 17,500 species per year, assuming 5 million species in the tropical moist forests (Wilson, 1988). The immediate, or proximal, causes of deforestation are easy to identify: the conversion of forest to agricultural use, logging, mining, industrial development (e.g., hydroelectric power), and the search for fuel wood and fodder. But what lies behind these proximal causes? A popular notion associates the destruction of tropical moist forests with population pressure. But the best analyses suggest that in some important forest regions, such as the Amazon, this argument explains, at most, only a small part of deforestation (see Chapter 3).

In fact, a constellation of distinct, though interacting, forces appears to be at work. The destruction of the tropical moist forest of Brazil's Amazon Basin offers an example. One key force manifests itself in the pressure of international markets for mineral and wood products, coupled with the country's interest in encouraging exports for the purpose of reducing its international debt. Public policies intended to promote economic development also contribute to the destruction of Amazonian forests (for example, government-sponsored road construction, hydroelectric power projects, and favorable tax treatment for large ranching operations, though the last has recently been withdrawn). Corporations cut trees to initiate mining and smelting operations, and more trees to fuel the processes. A frontier mentality encourages rapid use of forest resources on the assumption that there will always be new frontiers to open up when current areas are exhausted. And a history of nationalistic feelings—only now changing—encourages forest clearing: people have felt that Brazil must develop all of its resources to assume its rightful place as a world power. The incentives for policy makers to promote conservation have been weak: many of the costs of destroying tropical moist forests (for example, the resultant loss of biodiversity) are unknown or borne by those residing in other parts of the world, and consequently were ignored until recently by Brazilian policy makers. Under prevailing institutional arrangements, the incentives for individuals and corporations favored "mining" the forests, not conserving them. Although the proximal causes of the destruction of ancient forests are not difficult to pinpoint, the story becomes much more complex as we shift our attention to underlying causes that must be understood in order to control or redirect the behavior in question. The story is even more complex when one moves beyond Brazilian Amazonia and considers that the underlying causes and their relationships can be quite different from one country to another.

IMPLICATIONS

What do these stories tell us? They support the general notion that global environmental change is driven by trends in global production and consumption. But to recognize this is not necessarily useful for understanding specific types and rates of change or the complex set of factors influencing production and consumption. Single-factor explanations of the anthropogenic sources of global environmental change are apt to be misleading at best. It follows as well that simple, one-dimensional policies, such as a carbon tax or a uniform law of the atmosphere, cannot by themselves control global environmental change.

The stories also suggest the need to build stronger links between the natural sciences and the social sciences in efforts to understand global environmental changes and to devise public policies to respond to them in an effective manner. To project such changes, natural scientists must also project human behavior: what actions might affect the environment, how people might respond to environmental changes, and how people might use information in making decisions about their relationship to the environment. The quality of the environmental analysis is limited by the quality of the behavioral analysis that it includes. Erroneous assumptions about the future course of human behavior can lead analysts to incorrect projections of environmental change. Erroneous assumptions about how environmental change affects people can lead to neglect of feedback processes, including policy responses, that may mitigate or accelerate natural processes. Erroneous assumptions about human decision making can lead analysts to generate information that decision makers find useless or unresponsive to their needs.

Similarly, to understand the human dimensions of global change, social scientists must understand environmental processes. They need to understand what is known about how the greenhouse effect is likely to affect climate in order to understand which changes are likely to be noticed and thus lead to spontaneous responses. To evaluate alternative policy responses, they need to know which behaviors are believed to have major effects on the balance of atmospheric gases or the habitats of threatened species, and which, only minimal effects. Erroneous assumptions may lead social scientists to waste effort studying behaviors that are unlikely to occur or to have much impact on global change.

Consequently, effective cooperation between natural scientists and social scientists is essential for making progress. But such cooperation is not easy to achieve under the best of circumstances. We have more to say about this cooperation and about ways to strengthen it at a number of points in this book.

ORGANIZATION OF THE BOOK

This book articulates the principal elements of a strategy for adding to knowledge of the human dimensions of global environmental change. It takes stock of relevant existing knowledge, identifies what is unknown but might be learned about human behavior that could improve understanding of global change, and sets forth a series of programmatic guidelines to give direction to efforts over the next 5 to 10 years to improve that understanding. Chapter 2 presents a working definition of global change and a schematic model of the relationships between environmental and human systems. Chapters 3 and 4 elaborate on these relationships: Chapter 3 explores both the proximate and the underlying sources of anthropogenic change in large physical and biological systems. Chapter 4 concentrates in turn on human consequences of global environmental changes and responses to them. In these chapters, we outline current knowledge about the human dimensions of global change, identify important feedback loops between natural and human systems and between human causes of global change and human responses, and set forth an intellectual framework. We also identify particularly important near-term research needs whenever possible.

The report then turns to a number of problems that must be overcome in order to stimulate effective research. Chapter 5 examines a range of theoretical and methodological problems that confront efforts to broaden and deepen our understanding of the human dimensions of global change. Chapter 6 deals with data needs, including the needs to acquire baseline data on human systems, to assess and improve the quality of data, and to make relevant data more accessible to analysts. Chapter 7 takes up issues relating to human resources and organization that must be addressed in order to build a healthy program of research. These chapters include specific recommendations regarding research priorities; development of theory, methodology, and data; and improvement of human resources and the organizational basis for research. The chapters will be of particular interest to researchers and research managers concerned with providing an intellectual base for the long term. Chapter 5, in particular, will be of interest to social scientists who are looking for ways to relate global change research to progress in their disciplines.

In Chapter 8, we present the structure of a national research program on the human dimensions of global change.

2
Global Change and Social Science

Concern about global environmental change is sweeping through the scientific community, both in this country and abroad. Interest was organized at the international level through the International Geosphere-Biosphere Programme (IGBP), an activity of the International Council of Scientific Unions. The program officially focuses on issues of physical and biological science, although a few of the principals have emphasized the human dimensions of global change and sought to involve the social sciences (Clark, 1988; International Federation of Institutes for Advanced Study, 1987; Jacobson and Price, 1990; Kates, 1985a, b; Kates et al., 1985; for an important analysis by Americans, see Chen et al., 1983). More recently, natural scientists in the United States and some other countries have increasingly realized that global change cannot be understood, much less dealt with sensibly, in the absence of substantial contributions from the social sciences. For one thing, human responses to global change are likely to feed back into the processes at work to amplify, dampen, or redirect the changes in question. Even more important, the global changes of greatest interest today, like ozone depletion, climate change, and the loss of biodiversity, are largely anthropogenic in origin. Recognizing such phenomena, the National Research Council's Committee on Global Change, which originally consisted entirely of natural scientists, has consistently emphasized the importance of the human dimensions and has increasingly involved social scientists in its efforts (Clark, 1988; National Research Council, 1990b). Other interdisciplinary efforts sponsored by the International Federation of Institutes for Advanced Studies, the International Social Science Council, and other groups have also involved social scientists in the study of global change.

In this book, we use the term *social science* to cover a broad range of research activities usually associated with disciplines such as economics, sociology, political science, psychology, anthropology, geography, and history and interdisciplinary fields such as policy science, human ecology, and management. Social science involves the systematic study of the behavioral processes of individuals and social groups, organizations, and institutions. Our use of the term includes activities that are sometimes characterized as behavioral science.

Natural scientists interested in global change sometimes harbor false expectations about the contributions social scientists can make to understanding of global change. But the important point is that the U.S. global change community is now remarkably receptive to input from the social sciences. It is consequently time for social scientists to take seriously the challenge of mapping a strategy to add to knowledge of global environmental change.

GLOBAL CHANGE AND ENVIRONMENTAL SYSTEMS

Global environmental changes are alterations in natural (e.g., physical or biological) systems whose impacts are not and cannot be localized. Sometimes the changes in question involve small but dramatic alterations in systems that operate at the level of the whole earth, such as shifts in the mix of gases in the stratosphere or in levels of carbon dioxide and other greenhouse gases throughout the atmosphere. We speak of global change of this sort as *systemic* in nature because change initiated by actions anywhere on earth can directly affect events anywhere else on earth. Other times, the changes in question result from an accretion of localized changes in natural systems, such as loss of biological diversity through habitat destruction and changes in the boundaries of ecosystems resulting from deforestation, desertification or soil drying, and shifting patterns of human settlement. Global changes of this sort we describe as *cumulative* in nature; we consider them global because their effects are worldwide, even if the causes can be localized (for further exposition of the concepts of systemic and cumulative global change, see Turner et al., 1991b). The boundary between systemic and cumulative change is not sharp; it depends on how rapidly an environmental change spreads in space.

The most prominent global changes, as noted in Chapter 1, are increases in atmospheric greenhouse gases, depletion of the stratospheric ozone layer, and loss of biological diversity. There are others, however, such as pollution of the oceans (a systemic change) and possibly degradation of soil quality (a cumulative change)—and yet others are probably still unknown. People also debate whether the acidification of lakes and forests caused by the long-range transport of airborne pollutants is a global or only a regional change. Nevertheless, the appropriate strategy for scientific analysis of global change seems straightforward. One may conceive of the earth as a complex system composed of a number of differentiable but interacting spheres or subsystems. Some of these, including the atmosphere, the biosphere, the geosphere, and the hydrosphere, can be thought of as *environmental systems*, in that from the human perspective they constitute the environment. Others, sometimes called the noosphere or the anthroposphere and further subdivided—for example, into economic, political, cultural, and sociotechnical systems—can be distinguished as *human systems* (the terms *environmental systems* and *human systems* are taken from Clark, 1988). Approached in this way, the study of global change centers on efforts to understand how environmental systems at the global level affect or are affected by changes in any one of these spheres or subsystems. Key to this study is understanding the feedback mechanisms between subsystems that either amplify or dampen the initial impacts.

Much public concern with global change comes from the sense that amplifying (positive) feedback mechanisms involving environmental systems may be impossible to control once they get started. Accumulation of greenhouse gases, for example, may raise temperatures sufficiently to increase rainfall in the high north latitudes, threatening the capacity of plants and animals to survive in a rapidly changing environment (ARCS Workshop Steering Committee, 1990). These changes in turn may contribute to an acceleration of climate change by lowering the earth's albedo through reduction in snow cover and sea ice. Warming will also affect the global hydrological cycle by changing precipitation and evaporation patterns, leading to shifts in vegetative cover; these changes in turn could amplify the warming in areas where desertification is taking place. Another possibility is that some global changes may trigger dampening (negative) feedbacks that offset or even dominate the forces unleashed by positive feedback processes.

Warming may increase low-level sea clouds leading to cooling due to the radiative effect of increased planetary albedo. Increases in the level of carbon dioxide in the atmosphere, for example, may stimulate the growth of plants through leaf changes that increase photosynthetic activity and CO_2 uptake, thereby moving the planetary climate system back toward its initial condition.

There is nothing new about global change. The earth has always been a highly dynamic system whose atmospheric, biological, and geological properties have changed, sometimes dramatically, over the course of time. And there is nothing new about global change forcing humans to make drastic changes in their ways of life. Between about 12,000 and 25,000 years ago (and later, in Canada)—a recent time in the two-million-year existence of the human species—thick sheets of ice covered most of northern Europe and Canada. The sites of present-day New York and Paris had arctic climates, and sea levels were about 100 meters lower than they are today.

But the global environmental changes occurring now differ from those of the past in at least two ways that have profound consequences for our thinking about this subject. For one thing, the pace of global change has picked up dramatically. Methane concentration in the atmosphere has doubled in the past century; chlorofluorocarbons (CFCs), which accounted for about one-quarter of the anthropogenic contribution of greenhouse gases in the 1980s, were not present in the atmosphere before the 1930s (Houghton et al., 1990). Such changes require analysis on the time scale of centuries or even decades for understanding ozone depletion or global climate change. Equally important is the fact that the global changes we are concerned with today are largely anthropogenic in origin. Humans are no longer simply innocent victims compelled to adapt, in some cases rapidly, to large-scale changes in environmental systems resulting from forces beyond their control. Instead, it is human behavior itself that must be controlled if we are to succeed in ameliorating or redirecting global change.

Trends in Global Change

A remarkable feature of the current concern about global change is that it is largely anticipatory of any effect on humanity. None of the environmental changes in question has moved beyond the early stages of its projected trajectory, and several of the global changes of greatest concern have yet to manifest themselves in any unambiguous and convincing fashion. Given the short-term

nature of so much human activity, and considering the great uncertainties about the future course of global change, the level of
concern about global environmental change now expressed in a
variety of public forums is extraordinary. In the interest of providing a common vantage point from which to examine the human dimensions of global change, we present a brief synopsis of
current expectations in the scientific community regarding trends
in ozone depletion, climate change, and the extinction of species.

There is general agreement that the yearly reductions in ozone
over Antarctica are a reliable pattern (seasonal depletions of up to
50 percent were measured in 1987 and again in 1989 and 1990),
that emissions of CFCs into the atmosphere are a principal cause
of this phenomenon, and that CFCs already in the stratosphere
will constitute a growing source of ozone depletion for several
decades, regardless of efforts to reduce or eliminate additional
emissions. The effects of the resultant increase in ultraviolet
radiation (more specifically, UV-B) reaching the earth's surface,
moreover, are widely believed to include damage to human health
(in such forms as skin cancers, cataracts, and suppression of the
human immune response system) and to plants and aquatic organisms (including crops of considerable importance to humans). Beyond this, our knowledge of ozone depletion is less clear-cut. There
is some evidence of Arctic depletions that are significant, though
less severe, than those recorded in Antarctica; decreases of a few
percent are observed. Predictions of future trends in ozone depletion are sensitive to changes in a number of variables, including
human responses to the threat of severe ozone depletion. And
much remains to be learned about the consequences of increased
ultraviolet radiation (Solomon, 1990).

Global climate change is undoubtedly more complex than ozone
depletion, more difficult to project, and more important in terms
of its potential impacts on human welfare. Projections of temperature trends over the next century are based largely on scenarios of increasing concentrations of greenhouse gases in the
earth's atmosphere (chiefly carbon dioxide, methane, CFCs, and
nitrous oxides). Projections of the equilibrium temperature response, expressed as a global mean temperature, are shown in
Figure 2-1 for four scenarios, one assuming current growth rates
and the others assuming progressively increasing controls of greenhouse gas emissions to the atmosphere. The best available analytical tools project that, assuming current growth rates for emissions, we can expect a significant rise in worldwide equilibrium
temperature (perhaps 1-5 degrees Celsius by the middle of the

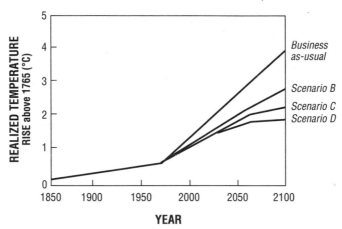

FIGURE 2-1 Best-estimate projected increases in global mean temperature due to observed (1850-1990) and projected emissions of greenhouse gases under business-as-usual assumptions and resulting from Intergovernmental Panel on Climate Change scenarios B, C, and D, which assume increasing levels of control of emissions. Uncertainty about the effect of emissions on temperature change could increase the 2100 temperature under business-as-usual assumptions by as much as an additional 2.0° C or decrease it by 1.4° C. Source: Houghton et al. (1990)

next century). However, the models are affected by several major uncertainties (for example, the role of clouds as reflectors of sunlight and of the oceans as carbon sinks), and sizable regional variations in temperature change are expected. In addition, there are sharp differences of opinion concerning the question of whether the trends predicted as a consequence of the greenhouse effect have already begun to appear in the actual data on global temperatures. While many observers interpret the increase of approximately 0.5 degree Celsius in mean global temperature recorded during the twentieth century as induced by the greenhouse effect, "it is not yet possible to attribute a specific portion of the . . . warming to an increase of greenhouse gases" (Folland et al., 1990:199). Predictions of climatic changes other than warming, such as in precipitation and cloud patterns, are even more uncertain (Houghton et al., 1990).

Though the total number of species on earth is not known, conservative estimates suggest a number of 3 to 10 million, of which approximately 1.4 million have been formally described (Wilson, 1988); recent estimates, however, range up to 30 or even 50 million species (Erwin, 1982, 1988). The history of species diversity on the planet appears to be one of long-term increases punctuated by episodes of mass extinction. Students of biodiversity recognize a number of mass-extinction events of varying magnitude, the one at the end of the Permian period being the most severe. The number of species appears to have increased since the mass extinction at the end of the Cretaceous period, 65 million years ago, but it has recently been declining at an unprecedented rate due to human activities. The rate of extinctions has been estimated at 1,000 or more times that before human intervention (Wilson, 1988). Destruction of land, fresh-water, and marine ecosystems is occurring worldwide; the destruction of tropical moist forests, which are thought to contain more than half of the earth's species, is the most important single cause of the acceleration in the extinction rate. The probable consequences of this trend for human welfare are not easy to foresee. The fact that humans now exploit only a tiny fraction of the earth's species may encourage a somewhat cavalier attitude toward the preservation of biological diversity. Yet there is a strong utilitarian case to be made for preserving the diversity of species on the grounds that future discoveries may demonstrate significant uses for many species and that species of no economic value may play critical roles in maintaining the stability of large ecosystems. In addition, there are powerful aesthetic and ethical arguments for maintaining biological diversity.

CHARACTERISTICS OF ENVIRONMENTAL SYSTEMS

Large physical and biological systems exhibit a number of characteristics that present challenges to those endeavoring to understand them. They require advances in scientific concepts, theories, and methods beyond those typical of existing disciplines. We briefly summarize these characteristics here. In Chapter 5, we note that human systems have analogous properties that pose very similar challenges for the social sciences.

Complex interdependencies exist both within and between environmental systems. Changes in one part of the earth's environment can have effects in surprising places. For instance, a recent proposal is that adding iron to the oceans may reduce the buildup of carbon dioxide in the atmosphere by removing the limiting factor to the growth of phytoplankton, which absorb excess carbon dioxide from the air (Martin et al., 1990). Although serious questions have been raised about the proposal (e.g., Lloyd, 1991), this sort of phenomenon makes predictions of environmental changes difficult. Causes and effects can be widely separated in space, and the knowledge necessary for prediction often requires contributions from scientific disciplines that do not ordinarily communicate with each other. Human activities intended to affect only one aspect of the environment can have far-flung and unanticipated consequences. Because no one can foresee when and how human activities will produce undesired effects, it is not clear in advance how to control them.

Global environmental systems frequently exhibit *nonlinear responses*. Mathematical models of global processes demonstrate that, under certain conditions, small perturbations in environmental systems can have large effects. In principle, a minute air current, of the sort a butterfly fluttering its wings might produce, could cause a major storm halfway around the world. Similarly, some small changes in human activities can produce huge effects—yet some large changes may make no difference. The net result is great uncertainty in predicting relationships between initial changes and final outcomes. Scientific analysis cannot easily come to terms with uncertainty of this sort because so little is known about the thresholds in either natural or human systems at which incremental changes are sufficient to trigger sharp discontinuities. Still, the phenomena are important because the most serious impacts of gradual environmental changes on human welfare (for example, the buildup of carbon dioxide) may result from an increased frequency of catastrophic events, like floods and crop failures, rather than from slow changes in average temperatures.

Environmental systems can undergo *irreversible changes*. The clearest example is the extinction of species. Ecosystems can also go extinct as a result of pollution beyond the point of no return or the conversion of their locales to human uses. Climate changes that cause forests to "migrate" can move them to locations from which they cannot naturally return, even if the climate system reverts to its original condition. Deforestation in some tropical areas causes soils to become unfit over time for annual crops or for revegetation by preexisting species. Discontinuities of this sort cause concern not only because of the value of what is lost but also because irreversible changes can reverberate through interdependent systems to cause additional changes that may be irreversible as well. And it is difficult to predict which environmental changes will have irreversible effects.

Long lag times are common in environmental systems. CFCs released into the atmosphere migrate to the stratosphere, where they are broken down by sunlight over a period of decades to several centuries. CFCs released in the 1990s will continue to destroy ozone in the stratosphere well into the twenty-first century and, in some instances, beyond. Because of such slow effects and the interdependencies of environmental systems, many human interventions in the global environment constitute uncontrolled experiments whose results may not be known for generations. This makes knowledge difficult to accumulate. It also increases the demand for knowledge both because these experiments may threaten the whole earth and because they have the potential to set catastrophes in motion before their effects are even noticed.

Global environmental changes can result from the *interactions of local systems* with each other and with larger-scale systems. For some analytic purposes, it is inadequate to treat the earth as possessing a single environment. Although the atmosphere is global, understanding of the biosphere may need to be built up from knowledge at smaller spatial scales, such as ecosystems or biomes. Thus, knowledge of global change requires ways to understand relationships across spatial scales (Clark, 1987; Rosswall et al., 1988). Human activities compound the challenge by redistributing species and transforming habitats, thus altering the ways ecosystems interact.

These characteristics of the global environment present serious challenges for scientific research and may call for new theories and methods. In addition to progress on scientific questions that fall within standard disciplinary boundaries, problems of global change require approaches that treat the earth as a single interactive system and stress the powerful interdependencies among environmental (and human) systems. Such approaches tend to be interdisciplinary rather than multidisciplinary (Schneider, 1988) and are often characterized by holistic analytic premises such as those of ecology or systems analysis.

The nature of the global environment also raises doubts about the value of the existing structure of scientific disciplines for understanding global change. To the extent that resources continue to be channeled through the familiar disciplines, the disciplines look increasingly like part of the problem. Those working on computer models of global climate change (that is, general circulation models) already find it necessary to incorporate into their algorithms variables and relationships from various disciplines of physical science; biological variables and relationships will increasingly be included as the models are refined. And pressure is growing to incorporate projections of human activities into evolving models of the earth system (National Research Council, 1990b:111). The need to understand global change may well become a powerful force for change in the existing structure of scientific disciplines.

ENVIRONMENTAL SYSTEMS AND HUMAN SYSTEMS

Research on the human dimensions of global change strives to understand the interactions between human systems and environmental systems, particularly global environmental systems, and to understand the aspects of human systems that affect those interactions. Human systems include economies, populations, cultures, governments, organizations that make technological choices, and so forth. Many of them are associated with disciplines that specialize in their study. Environmental systems include systems of atmospheric gas exchange, biogeochemical dynamics, ocean circulation, ecological interactions of populations of organisms, and so forth. These also tend to be associated with academic specializations.

Interactions between human systems and environmental systems have two critical interfaces, as shown in Figure 2-2. One is the subset of human actions that act as proximate causes of environmental change, that is, that directly alter aspects of the environment. The other is the subset of outcomes of environmental systems that proximally affect what humans value.

The example of anthropogenic climate change can clarify the relationships involved. Each human system has its own internal dynamics, and each also interacts with other human systems and the environment. Some of the activities of human systems, such as fossil fuel burning and agricultural conversion of wetlands, are significant proximate causes of global environmental change. That is, they directly alter aspects of the environment in ways that have global effects. These particular proximal causes add carbon dioxide and methane to the atmosphere and thus contribute to the greenhouse effect. The human causes of global environmental change, which are the focus of Chapter 3, include the human activities that proximally, or directly, alter the global environment and the aspects of human systems that explain those activities and therefore affect the global environment indirectly through their effects on the proximal causes. It is important to emphasize that the human causes of global environmental change quite often depend on decisions made and actions taken without any consideration of the global environment.

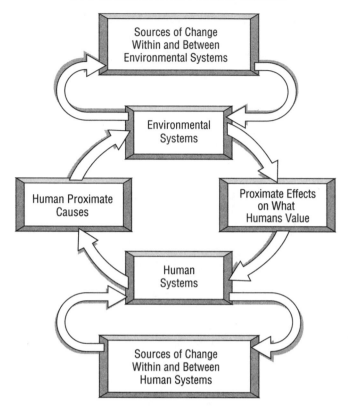

FIGURE 2-2 Interactions between human and environmental systems. Source: This figure elaborates on Clark (1988:Figure 1).

Environmental systems, like human systems, have their own internal dynamics. And like human systems, each environmental system interacts with other systems, both environmental and human. Some of the processes in environmental systems, whether or not human systems were involved in causing them, proximally affect things humans value. For example, processes that warm the atmosphere might result in rainfall patterns that inhibit or enhance the growth of crop plants or dry up sources of surface water used for human consumption. We use the phrase, "what humans value," broadly. It refers not only to outcomes that affect human health and material well-being, but also to outcomes such as extinction of species, disruption of ecosystems, and loss of natural beauty, on which humans may place aesthetic, spiritual, or intrinsic value. Chapter 4 focuses on these consequences of global change—how global change may affect what humans value and how those effects, or the anticipation of them, may affect human behavior.

QUESTIONS FOR NATURAL SCIENCE, QUESTIONS FOR SOCIAL SCIENCE

The human causes and consequences of global change raise questions for both natural and social science. On the causes side, important human actions include releasing CFCs, burning fossil fuels, and cutting tropical moist forests. Much remains to be learned about the effects of these actions on the global environment, with many puzzles for natural scientists to solve. We do not yet understand, for example, where all the carbon dioxide emissions go, or how large an area certain ecological communities need to remain viable. The principal issues for the social sciences center on the causes of the human actions that proximally cause global change. On the consequences side, natural scientists need to address such questions as the effects of ozone depletion on the incidence of certain types of cancer and the implications of global climate change for agricultural production. The key issues for social scientists center on human responses to actual or anticipated global changes. In addition, as Figure 2-2 shows, human responses to actual or anticipated global changes frequently trigger feedback processes that affect the anthropogenic sources of global change. For example, faced with the prospect of ozone depletion or global warming, humans may act to reduce or eliminate their consumption of CFCs or their use of fossil fuels or make changes in demographic patterns or institutional arrangements. Such possibilities also fall into the domain of social science. We address them in Chapter 4.

CONTRIBUTIONS FROM SOCIAL SCIENCE

What can the social sciences contribute to understanding the human dimensions of global change during the next decade? We have approached this question in two distinct ways. One is to imagine answering queries from policy makers and natural scientists working on global change issues. The other is to identify broad conceptual and theoretical constructs from social science that could be brought to bear on the problem in an illuminating fashion.

Policy makers and natural scientists interested in global change are likely to ask: Why does the United States consume so much more energy per unit of gross national product than most other Western industrialized countries? Why are ancient forests destroyed more rapidly in some societal settings than in others? How can we account for large discrepancies in land use practices, even among societies that resemble each other in many ways? These are complex questions that social science cannot answer now with confidence. Nonetheless, such questions can be analyzed with social science techniques, and a serious literature relating to many questions of this sort is available for study.

For example, researchers who examine the different patterns of energy consumption in the United States and Canada on the one hand and Western Europe and Japan on the other have much to say about the relative importance of geography (for example, distances between human settlements), demography (for example, the dispersal of human populations into suburbs), economics (for example, the relative costs of labor and energy as factors of production), infrastructure (for example, the prevalence of central heating in homes), and public policy (for example, taxes, subsidies, and policies governing rents from natural resources) as determinants of the propensity of North Americans to rely more heavily on energy as a factor of production than Europeans or especially the Japanese do (e.g., Schipper et al., 1985; Schipper, 1989). Similarly, researchers who examine the pace of deforestation and the spread of large-scale cattle ranching in the Amazon Basin are engaged in a lively debate about the relative importance of institutional factors (for example, systems of land tenure), technological factors (for example, the introduction of modern road building and land clearing equipment), international economic factors (for example, the growth of a world market for lean beef), political factors (for example, various forms of tax relief and public subsidies for activities involving land clearing), cultural factors (for example, the tendency to apply a frontier mentality to decisions about the Amazon), and population growth as determinants of the destruction of Amazonian moist forests (e.g., Hecht, 1985; World Resources Institute, 1985). The techniques exist to greatly improve understanding of how such factors act separately and together to influence the proximate human causes of global change.

The social sciences can also contribute by using available conceptual and theoretical constructs to illuminate problems of global change. For example, one of the most powerful and well-documented findings in social science is that apparently rational actors engaged in interactive decision making can and often do end up with outcomes that are less than optimal—and in some cases highly destructive—for all concerned (e.g., Hardin, 1982). Such problems of collective action as they have come to be known, arise regularly in conjunction with the use and abuse of open-access resources such as the stratospheric ozone layer, the earth's climate system, and the planet's gene pool, as well as with the supply of public or collective goods, such as air and water quality. It follows that much might be learned about the human causes and consequences of global change from research on collective-action problems and on possible solutions for them, especially if the research focused on global change.

The same is true for research on "social traps," situations in which actions that are initially rewarded or reinforced lead to behavior or habits with later consequences that those involved would rather avoid (Cross and Guyer, 1980). Mundane examples of traps include smoking and drinking, habitual behaviors that addicts often find difficult or impossible to change even when the painful consequences become abundantly clear. On a much larger scale, social traps are an analogue for the problems of controlling or redirecting the anthropogenic sources of global change. For example, American society is "addicted" to the intensive use of energy in the sense that cheap energy has changed society in ways that make it increasingly difficult to return to past habits. Energy-dependent patterns of dispersed, suburbanized settlement make it difficult to adopt energy-efficient technologies such as mass transit that were appropriate in the more densely packed cities of the past.

The examples of collective-action problems and social traps suggest that the selective development of some fields of traditional social science can help illuminate processes affecting global change. If researchers in these fields are encouraged to focus on cases relevant to the global environment, they may improve fundamental understanding of global change. This theory-based research strategy has the potential, over time, to enable social science to address more confidently the pointed, policy-oriented questions that will continue to arise.

KNOWLEDGE BASE OF ENVIRONMENTAL SOCIAL SCIENCE

Most of the social science research relevant to global environmental change has been undertaken and organized not so much within the traditional disciplines as within subfields that are usually interdisciplinary or multidisciplinary in scope, such as cultural ecology, environmental history, environmental perception, human ecology, natural hazards research, resource economics, and resource management, to name a few. These research areas taken together constitute a broader field of environmental social science, a cluster of research activities that takes human–environment relationships as its focus, including human-induced physical changes, perceptions of environmental changes from whatever cause, and responses to the environment. Most of the research has been conducted below the global scale. We briefly sketch the outlines of the domain of environmental social science, citing some sources of more detailed information on recent research activities.

Many of the subfields that constitute environmental social science tend to emphasize human reactions to the environment—perceptions and responses. For example, *environmental perception* studies, blending geography and psychology, have focused on the structures and processes of human learning and cognition as people interact with their surroundings, whether natural or social (Aitken et al., 1989; Fischhoff and Furby, 1983; Whyte, 1985; Golledge, 1987). Studies in *environmental sociology* and *political science* have examined, among other issues, the implications of social movements (Morrison, 1991), public opinion (Dunlap and Jones, 1991), and political economy (Schnaiberg, 1991) for human responses to environmental problems (see also Buttel, 1987; Heathcoate, 1985). Research in *environmental psychology* has addressed human responses to environmental stressors (Baum and Paulus, 1987; Evans and Cohen, 1987; Fischhoff et al., 1987), as well as the determinants and ways of altering individual behaviors that affect the natural environment (Stern and Oskamp, 1987). *Natural hazards* studies have attempted to create global and cultural typologies of perceived environmental hazards and impacts, and of human responses to them (Burton et al., 1978; Mitchell, 1989; Mileti and Nigg, 1991). Of all the environmental social sciences, this subfield has the longest tradition of research that is global in scope (e.g., Kotlyakov et al., 1988; Parry et al., 1988). The emphasis on the "built environment" in the environment and behavior literature in environmental psychology and environmental sociology makes this research tangential to some issues of global change; nevertheless, these subfields provide empirical, methodological, and theoretical insights useful to some aspects of the problem (e.g., Dunlap and Michelson, 1991; Craik and Feimer, 1987).

Another group of environmental social sciences has emphasized human activity as an influence on the physical environment, primarily through examinations of the transformation of the physical landscape and the societal forces that give rise to it. *Cultural* (also political) *ecology, human ecology,* and *environmental history* have focused on the character of the human–environment relationship per se, primarily through examinations of the transformation of the physical landscape and the societal forces that give rise to it. This tradition of nature–society studies can be traced at least to the seminal work of George Perkins Marsh (1864), through the 1955 symposium on "Man's Role in Changing the Face of the Earth" (Thomas, 1956), and a more recent effort on "The Earth as Transformed by Human Action" (Turner et al., 1991a). Save for the last work, these and similar efforts have typically focused on regional and local relationships because of the importance of context in understanding the forces that drive changes and the human adjustments to them (Butzer, 1989; Ellen, 1982; Steward, 1955). Attention has been given both to long sweeps of prehistory and history (e.g., Butzer, 1976; Cronon, 1983; McEvoy, 1986; Merchant, 1991; Pync, 1982; Rabb, 1983; Richards, 1986), and to contemporary change (e.g., Blaikie and Brookfield, 1987; Netting, 1968; Rappaport, 1967; Turner and Brush, 1987; Rosa et al., 1988). Human ecology incorporates these regional, local, and historical concerns but also attempts to integrate micro-social phenomena and social interactions into an understanding of human–environment relations and draws on evolutionary approaches to social and environmental change (Borden et al., 1988; Dietz and Burns, 1991; Dietz et al., 1990). Such studies have typically demonstrated the complexity of human–environment relationships and the significant degree to which presumed broader forces are mediated by local socioeconomic and environmental conditions.

Resource economics is somewhat unusual in environmental social science in representing an approach clearly identified with a single discipline, though now of much wider currency. From Malthus to the present, a central theme in resource economics has been natural resource depletion induced by growth, human population, and the economy, and the threat this poses to human welfare (Barnett and Morse, 1963; Smith, 1979). More recently, concomitant with the attempt to develop a field of ecological economics, the depletion theme has been extended to include environmental resources, including those "provided" by ecosystems and the atmosphere, and the spatial management problems these pose because of the absence or weakness of markets for exchanging them. Emphasis is on the magnitude and spatial and temporal distribution of the social costs and benefits of using these resources. Analysts propose institutional arrangements that would internalize the costs and benefits. Some favor publicly administered regulations and penalties; others favor the use of more market-like incentives, such as taxes or subsidies, to achieve the same end (Kneese and Russell, 1987). *Resource management*, an interdisciplinary subfield that combines physical and social science, focuses on physical resources such as water, oil, or wildlife. The field draws on legal and social theory so that its management concepts take into account objectives other than growth or efficiency, such as social cohesion or preservation of cultural groups (Emel and Peet, 1989; Heathcoate and Mabbutt, 1988; Rees, 1985; Savory, 1988).

SETTING PRIORITIES FOR SOCIAL SCIENCE RESEARCH

Social science has potential for contributing to knowledge about global change, starting from either practical questions or relevant theory. This is good news for social scientists because it indicates that global change research can become a rich new field of study. But it also means that priorities must be set to effectively allocate the scarce resources of time and money. In later chapters, we review current knowledge and address the issue of priorities for a national research program on the human dimensions of global change. Among the research needs identified there, highest priority should be given to research activities that meet several of the following criteria.

1. Importance. The research addresses fundamental issues relating to the anthropogenic sources of global environmental change or the human consequences of or responses to global change. We favor studies that seem likely to shed light on patterns of behavior that:

—have high impact, for example, if a proximate cause of global change, such as emissions of major greenhouse gases, is highly sensitive to the behavior or if a behavior is highly sensitive to an anticipated global change;

—have multiple impacts, for example, increases in crop acreage have combined effects on biological diversity and releases of carbon dioxide, nitrous oxide, and methane;

—are likely to become more important over a time scale of decades or centuries, even if they do not have high or multiple impacts now;

—affect the robustness of societies under the impact of global environmental changes, that is, their ability to withstand such changes without major disruptions of human life;

—affect the capacity to respond appropriately to global change, for example, by promoting or impeding wise use of scientific knowledge or facilitating or inhibiting agreements on response between political units;

—may have irreversible effects on the global environment or on people's ability to respond to environmental change.

2. Relevance for Action. The results of the research have potential value to individuals and organizations, including government agencies, that take action in response to global change. It is a mistake to take a narrow view of the criterion of relevance. Some of the most striking developments in the social sciences (for example, the theory of collective action) as well as in the natural sciences have originated in basic research whose initial links to applied concerns were anything but obvious. There is a need for research probing the roots of the anthropogenic sources of global change, and not only research that seeks answers to specific practical questions. Still, we believe it is important to give priority to studies that are relevant in the sense that they promise to shed light on decision variables that are actually or potentially subject to human control.

3. Improving Theory. The research helps develop theoretical tools that will facilitate future studies of the human dimensions of global change. Among the most general theoretical needs are improved ways to analyze social change on the time scale of decades to centuries and to make connections between different levels of analysis and of spatial and temporal aggregation. (These points are elaborated further in Chapter 5.)

4. Adding to Existing Knowledge. The research has a high probability of making a large marginal contribution to knowledge. It may deal with an issue about which relatively little is known or one about which available knowledge is highly uncertain; there should be a strong probability of a useful result. This criterion may be met in various ways: the research might address important practical questions, such as the relative effectiveness of different kinds of interventions in response to global change; timeliness is an important additional criterion for such research. It might develop knowledge on critical broad problems, such as the nature of interactions among the driving forces of global change or the functioning of decision or conflict processes in responses to global change. Or it might help develop theory. There is much to be said, for example, for supporting research in areas in which extensive data sets coexist with relatively underdeveloped conceptual and theoretical foundations. An example may be research efforts to explain trends in the consumption of fossil fuels in

various parts of the world. As we note in subsequent chapters, many research methods are useful, and the program of research on the human dimensions of global change should be characterized by methodological pluralism.

5. Improving Data.   The research helps raise the quality of data on important variables relevant to the human dimensions of global change, improves access to such data, or provides data on important variables for which good measures do not yet exist.

6. Amenability to Research.   The research uses established techniques of social science or appropriate techniques newly developed for global change research.   Some questions lend themselves more easily than others to research by social scientists because of the existence of relatively large universes of comparable cases, the ease of operationalizing key variables, or the usefulness of computer simulation as a modeling technique. (We discuss these issues in greater detail in Chapter 5.)

7. Interdisciplinary Potential.   The research has strong potential to contribute to effective communication across social science disciplines, or between social scientists and natural scientists working on global environmental change, and to facilitate collaboration that bridges intellectual divisions.   It seems evident that success in understanding global change will require effective alliances across disciplines, particularly between social scientists and natural scientists.   Yet real successes in forging such alliances have been few and far between, despite frequent declarations concerning the importance of interdisciplinary studies.   We therefore take the view that priority should be given to studies that offer imaginative ways to solve this problem.

8. Potential for International Collaboration.   Research that meets the other criteria should be given higher priority if, by its organization or its likely products, it can be expected to strengthen the ability of an international community of researchers to gain understanding of the human dimensions of global change.

CONCLUSION

The global environmental changes of greatest current concern are inextricably intertwined with human behavior.   They cannot be understood without understanding the human activities that cause them and the ways humans may respond to the awareness of global change.   This state of affairs dictates that social scientists apply their knowledge and methods to the problem of global change and that natural and social scientists work together to build the needed knowledge of how human and environmental systems interact. Although environmental social science can contribute by directly analyzing policy questions, in the long run it is critically important to build theory, methods, and data bases that can improve environmental social science. In that way, basic understanding of the major types of decisions and behaviors that cause or respond to global change can also grow.

3

Human Causes of Global Change

All the human causes of global environmental change happen through a subset of proximate causes, which directly alter aspects of the environment in ways that have global effects. We begin this chapter by outlining and illustrating an approach to accounting for the major proximate causes of global change, and then proceed to the more difficult issue of explaining them. Three case studies illustrate the various ways human actions can contribute to global change and provide concrete background for the more theoretical discussion that follows. We have identified specific research needs throughout that discussion. We conclude by stating some principles that follow from current knowledge and some implications for research.

IDENTIFYING THE MAJOR PROXIMATE CAUSES

The important proximate human causes of global change are those with enough impact to significantly alter properties of the global environment of potential concern to humanity. The global environmental properties now of greatest concern include the radiative balance of the earth, the number of living species, and the influx of ultraviolet (UV-B) radiation to the earth's surface (see also National Research Council, 1990b). In the future, however, the properties of concern to humanity are likely to change—ultraviolet radiation, after all, has been of global concern only since the 1960s. Consequently, researchers need a general system for moving from a concern with important changes in the environment to the identification of the human activities that most seriously affect those changes. This section describes an accounting system that can help to perform the task and illustrates it with a rough and partial accounting of the human causes of global climate change.

A Tree-Structured Accounting System

A useful accounting system for the human causes of global change has a tree structure in which properties of the global environment are linked to the major human activities that alter them, and in which the activities are divided in turn into their constituent parts or influences. Such an accounting system is helpful for social science because, by beginning with variables known to be important to global environmental change, it anchors the study of human activities to the natural environment and imposes a criterion of impact on the consideration of research directions (see also Clark, 1988). This is important because it can direct the attention of social scientists to the study of the activities with strong impacts on global change.

Because the connections between global environmental change and the concepts of social science are rarely obvious, social scientists who begin with important concepts in their fields have often directed their attention to low-impact human activities (see Stern and Oskamp, 1987, for elaboration). An analysis anchored in the critical physical or biological phenomena can identify research traditions whose relevance to the study of environmental change might otherwise be overlooked. For example, an examination of the actors and decisions with the greatest impact on energy use, air pollution, and solid waste generation showed that, by an impact criterion, studies of the determinants of daily behavior had much less potential to yield useful knowledge than studies of household and corporate investment decisions or of organizational routines in the context of energy use and waste management (Stern and Gardner, 1981a,b). Theories and methods existed for each subject matter in relevant disciplines such as psychology and sociology, but much of the research attention had been misdirected.

The idea of tree-structured accounting can be illustrated by the following sketch of a tree describing the causes of global climate change.

1. The chief environmental property of concern is the level of greenhouse gases in the atmosphere. The major anthropogenic

greenhouse gases, defined in terms of overall impact (amount in the atmosphere times impact per molecule integrated over time), are carbon dioxide (CO_2), chlorofluorocarbons (CFCs), methane (CH_4), and nitrous oxide (N_2O). If the trunk of the tree represents the greenhouse gas-producing effect of all human activities, the limbs can represent the contributing greenhouse gases. Table 3-1 presents the limbs during two different time periods and a projection for a future period.

2. Both natural processes and human activities result in emissions of greenhouse gases. For instance, carbon dioxide is emitted by respiration of animals and plants, burning of biomass, burning of fossil fuels, and so forth. If each limb of the tree represents human contributions to global emissions of a greenhouse gas, the branches off the limbs can represent the major anthropogenic sources of a gas, that is, the major categories of human activity that release it. These are proximate human causes of climate change, and their impact is equal to their contribution of each greenhouse gas times the gas's radiative effect, integrated over time. For the same emissions, the representation of impact will vary with the date to which the impact is projected. Tables 3-2 and 3-3 allocate emissions of the most important greenhouse gases during the late 1980s to human activities.

3. Major human proximate causes, such as fossil fuel burning, are conducted by many actors and for many purposes: electricity generation, motorized transport, space conditioning, industrial process heat, and so forth. A tree branch, such as one representing fossil fuel burning, can be divided into twigs that represent these different actors or purposes, each of which acts as a subsidiary proximate cause, producing a proportion of the total emissions. It is possible to make such a division in numerous ways. Fossil fuel burning can be subdivided according to parts of the world (countries, developed and less-developed world regions, etc.), sectors of an economy (transportation, industrial, etc.), purposes (locomotion, space heating, etc.), types of actor (households, firms, governments), types of decisions determining the activity (design, purchase, utilization of equipment), or in other ways. Different methods may prove useful for different purposes. Table 3-4 illustrates one way to allocate the carbon dioxide emitted from fossil fuel consumption to the major purposes (end uses) of those fuels.

4. The tree structure can be elaborated further by dividing the subsidiary proximate causes defined at the previous level into their components. Such analysis is important for high-impact activities.

TABLE 3-1 Estimated Human Contributions Per Decade to Global Warming of Major Greenhouse Gases During Three Time Periods, in Watts per square meter (percentage in parentheses)

Gas	1765-1960[a]	1980s[a]	2025-2050 Projection[b]
CO_2	0.059 (68)	0.30 (55)	0.51 (67)
CH_4	0.018 (21)	0.06 (11)	0.07 (9)
CFCs, HCFCs	0.001 (1)	0.13 (25)	0.11 (15)
N_2O	0.003 (4)	0.03 (6)	0.04 (5)
Stratospheric H_2O[c]	0.006 (7)	0.02 (4)	0.024 (3)
Total	0.087	0.54	0.76

These estimates are of "radiative forcing" by greenhouse gases, that is, the change they produce in the earth's radiative balance that in turn changes global temperature and climate. Radiative forcing is calculated from current gas concentrations in the atmosphere, which include gases remaining in the atmosphere from all emissions since the beginning of the industrial era, set here at 1765. It is not identical to the "global warming potential" of gases emitted by human activity, a property that integrates the effects of gas emissions over future time. Global warming potential is affected by the different atmospheric lifetimes of greenhouse gases before breakdown, so that the relative importance of gases for global warming depends on the future date to which effects are estimated. In addition, chemical reactions in the atmosphere convert some radiationally inactive compounds into greenhouse gases over time. The estimation of the global warming potential of currently emitted gases is quite uncertain due to incomplete knowledge of the relevant atmospheric chemistry. An early estimate of the 100-year global warming potential of gas emissions in 1990 allocates it as follows: CO_2, 61%; CH_4, 15%; CFCs, 12%; N_2O, 4%; other gases (NO_x, nonmethane hydrocarbons, carbon monoxide), 8% (Shine et al., 1990). Although these estimates differ from the radiative forcing estimates in the table, the differences are not great in terms of the relative importance of the gases for the global warming phenomenon. Our analysis uses the estimates of radiative forcing because they are far less uncertain.

[a]Source: Shine et al. (1990:Table 2.6).

[b]Source: Shine et al. (1990:Table 2.7), assuming a "business-as-usual" scenario with a coal-intensive energy supply, continued deforestation and associated emissions, and partial control of CO and CFC emissions.

Uncertainties for the future projections are very large. Total effects of greenhouse gases projected for 2025-2050 varied by a factor of 5 from the "accelerated policies" scenario, which projected the lowest level of emissions, to the "business-as-usual" scenario, which projected the highest.

[c]Stratospheric water vapor is believed to increase as an indirect effect of CH_4 emissions.

TABLE 3-2 Global Emissions of CO_2, CH_4, and N_2O From Human Activities in the Late 1980s

Activity	Emissions	(%)	Range	Notes
CO_2 emissions (Mt carbon per year)				
Fossil fuel burning	5,400	(77)	4,900-5,900	
Tropical deforestation	1,600	(23)	600-2,600	
Total	7,000			
CH_4 emissions (Mt CH_4 per year)				
Rice paddies	110	(31)	25-170	Function of acreage and cropping intensity
Digestion in ruminants	80	(23)	65-100	Primarily domestic
Gas, coal production	80	(23)	44-100	
Landfills	40	(11)	20-70	Decay of wastes
Tropical deforestation	40	(11)	20-80	Biomass burning
Total	350			
N_2O emissions (Mt N_2O per year)[a]				
Fertilizer use	1.5	(38)		
Fossil fuel combustion	1	(25)		
Tropical deforestation	0.5	(13)		
Increased cultivation of land	0.4	(10)		
Agricultural wastes	0.4	(10)		
Fuel wood and industrial biomass	0.2	(5)		
Total	4			

Note: Mt = million metric tons

[a]Estimates of N_2O emissions are highly uncertain. For example, Watson et al. (1990) give a range of 0.01-2.2 for fertilization. In addition, N_2O releases from unknown sources are probably larger than all anthropogenic releases. It is not clear how much of the unaccounted releases is anthropogenic.

Sources: For CO_2 and CH_4, Watson et al. (1990); for N_2O, National Academy of Sciences (1991a).

For instance, automobile fuel consumption can be analyzed as the product of number of automobiles, average fuel efficiency of automobiles, and miles driven per automobile; the determinants of each of these factors can be studied separately. Researchers might then investigate the social factors that affect change in the number of automobiles and their typical life span, such as household income, household size, number employed per household, and availability of public transportation. More detailed analysis can be carried out until it no longer would provide information of high enough impact to meet some preset criterion. Again, there are many ways to analyze an activity such as automobile fuel consumption, and the most useful approach is not obvious a priori.

The task of making such accounts, even for a single tree, is enormous. The work can be eased by using the impact criterion: analysts might reasonably choose to move from trunk to limb to branch to twig only until the contribution falls below a preset level of impact for the time period of concern. Data collection and substantive analysis of the thinnest twiglets can be deferred. Table 3-5 presents a composite of the accounts of individual green-

TABLE 3-3 Anthropogenic Sources of Atmospherically Important Halocarbons in the Late 1980s

Halocarbon	Production × 10^6 kg/yr	Global Warming Potential[a]	Percent of Total Effect[b]	Uses[c]
CFC 11 (CCl_3F)	350	3,500	17	Aerosols, re-frigeration, foams
CFC 12 (CCl_2F_2)	450	7,300	60	Aerosols, re-frigeration, foams
CFC 113	150	4,200	13	Cleaning electronic components
HCFC 22 ($CHCl_2F$)	140	1,500	3	Refrigera-tion, poly-mers
CH_3CCl_3	545	100	2	Industrial degreasing
Others			5	

Note: Production estimates are from Watson et al. (1990), except for CH_3CCl_3, which comes from World Meteorological Organization (1985). Projections of future production are very sensitive to changes in economic growth, and relatively quick substitution is possible when alternative chemicals become available. CFC 22 production doubled between 1977 and 1984 (e.g., fast-food packaging), as did CFC 113 production (electronics industry).

[a]Numbers represent the integrated effects over 100 years of release of one unit mass of the compound, relative to CO_2. Integration over other time horizons would change the relative potentials because of differing atmospheric residence times. Source: Shine et al. (1990:Table 2.8).

[b]Percentage of 100-year effects of all 1990 halocarbon emissions. Source: Shine et al. (1990:Table 2-9).

[c]Projected atmospheric effects depend not only on total production but also on the balance between end uses. When CFC 11 and CFC 12 production shifted from aerosols to other applications after 1976, the result was a longer lag time from production to entry into the atmosphere.

TABLE 3-4 Disaggregation of Carbon Dioxide Emissions by Economic Sector and End Use (percentages, United States, 1987)

| End Use | Economic Sector | | | |
	Industrial	Buildings	Transportation	Total
Steam power, motors, appliances	19	7		26
Personal transportation (automobiles, light trucks)			20	20
Space heating	1[a]	16		17
Freight transport (heavy truck, rail, ship, other)			7	7
Heating for industrial processes	6			6
Lighting	1[a]	5		6
Cooling	—[a]	5		5
Air transportation			5	5
Water heating		3		3
Other	5			5
Total	32	36	32	100

Note: U.S. data are unrepresenative of world energy use in various ways. However, the United States is responsible for approximately 20 percent of global CO_2 emissions.

[a]2 percent in the single category of heating, ventilating, air conditioning, and lighting was allocated one percent each to heating and lighting.

Source: U.S. Office of Technology Assessment, 1991.

TABLE 3-5 Estimated Composite Relative Contributions of Human Activities to Greenhouse Warming

| Activity | Gases (Relative Contribution in percent) | | | | | |
	CO_2	CH_4	CFCs	N_2O	Other	Total
Fossil fuel use	42	3		1.5		46.5
CFC use			25			25
Biomass burn	13	1		1		15
Paddy rice		3				3
Cattle		3				3
Nitrogen fertilization				2		2
Landfills		1				1
Other				1.5	4	5.5
Total	55	11	25	6	4	101

Source: Compiled from Tables 3-1, 3-2, and 3-3. For interpretation of the data, see the note at Table 3-1.

house gases that gives the approximate contribution of major classes of human activity to climate change. Figure 3-1, one of the possible tree diagrams incorporating this information, identifies the human activities that call for finer-grained analysis on the basis of their impact. The more detailed, U.S.-centered accounting in Table 3-4 shows why much more detailed analysis is warranted for explaining the purchase and use of automobiles and light trucks with different levels of energy efficiency (per-

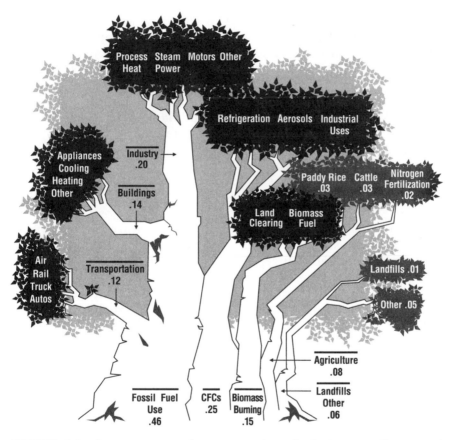

FIGURE 3-1 A tree-structured representation of relative contributions of human activities in the late 1980s to greenhouse warming. Note: Thicknesses of limbs and branches are proportional, where numbers are provided, to the contribution of the activity named. Where numbers are not provided, worldwide data were not available for further disaggregation. Even where numbers are provided, they are subject to varying degrees of uncertainty, as noted in the text. Sources: Table 3-5, except for the disaggregation of fossil fuel use into economic sectors, which was calculated from U.S. Office of Technology Assessment, 1991, Table 9-1 and Figure 9-2.

sonal transportation) than for explaining the choice or operation of water heating systems for buildings. For a policy-oriented analysis based on such an approach, see National Academy of Sciences, 1991b.

Accountings such as the one represented in Figure 3-1 can help guide the research agenda for the human causes of global change. They are critically dependent, however, on analyses from the natural sciences to sketch the trunk and major limbs, that is, to identify the most important environmental effects of human action and the technologies that produce those effects. Natural science can help social science by providing an improved picture of the trunk and limbs, and particularly by improving estimates of the uncertainties of their sizes. The uncertainties of some components are quite large (see, for instance, Table 3-2 estimating the relative contributions of different human activities to methane releases), and attention should be paid to whether, in the full account, these uncertainties compound or cancel each other. Research that estimates the relative impacts of proximate human causes of global change on particular environmental changes of concern, specifying the uncertainty of the estimates, is essential for understanding the human dimensions of global change.

As tree diagrams move from the trunk out toward the branches and twigs, analysis depends more on social science. For each important environmental change, there are several possible accounting trees, each consistent with the data but highlighting different aspects of the human contribution. Social science knowledge is needed to choose accounting procedures to suit specific analytic purposes. Whatever accounting system is used, social scientists conducting research on the human causes of global change should focus their attention on factors that are significant contributors to an important global environmental change.

LIMITATIONS OF TREE-STRUCTURED ACCOUNTING

Because many different tree diagrams may be consistent with the same data, tree diagrams must be treated as having only heuristic, not explanatory, value. They are useful but not definitive accounts. A more serious limitation of tree-structured accounts is that they do not by themselves illuminate the driving forces behind the proximal causes of global change. Social forces that have only indirect effects on the global environment, and that may therefore be omitted from tree accounts, can have at least as much impact as the direct effects. Consider, for instance, the rate of female labor force participation, which affects energy use in many different ways. With an increase in the proportion of women in the labor force, there tend to be more automobiles and miles driven per household, increased travel by plane, and, because of the associated decrease in household size, increased per capita demand for residential space conditioning and household appliances (see Schipper et al., 1989). Because these factors appear in different branches of Figure 3-1, the figure is not useful for representing the effect of female labor force participation on energy demand. The broader social process—the changing role of women in many societies—has even wider effects on energy use, but is still harder to capture in the figure. Despite these limitations, the accounting tree is useful as a preliminary check on the likely impact of a major social variable. When such a variable has a high impact, it is worth considering for inclusion in models of the relevant proximal causes of global change.

Tree-structured accounting is also limited in that it can evaluate human activities against only some criteria of importance (such as high and widespread impact), but not others (such as irreversibility). Consideration of criteria of importance other than current impact may require detailed empirical analyses of factors that look small in an accounting of current human causes of environmental change. An example, elaborated in the next section, concerns future CO_2 emissions from China. If per capita income grows rapidly there, Chinese emissions may increase enough to become tremendously important on a world scale. To make projections, it would be very useful to have detailed studies of the effects on emissions of increased income in other countries that have undergone recent spurts of economic growth, such as Taiwan and South Korea, even though these countries have no major impact on the global carbon dioxide balance.

EXPLAINING THE PROXIMATE CAUSES: THREE CASES

As we have shown, all human activity potentially contributes, directly or indirectly, to the proximate causes of global change. This section presents three rather detailed cases of human action with high impact on important global environmental changes to explore what lies behind the proximate causes. Taken together, the cases illustrate human causes that operate through both industrial and land use activities and in both developing and developed countries. They illustrate how multiple driving forces interact to determine the proximate human causes of global change and why systematic social analysis is necessary for understanding how human actions cause it. In the section that follows, we discuss the interrelationships among the driving forces at a more theoretical level.

THE AMERICAN REFRIGERATION INDUSTRY

In 1985, the head of the British Antarctic Survey, Joseph Farman, reported that his team had discovered a heretofore unobserved atmospheric phenomenon: a sudden springtime thinning of the ozone layer over Antarctica, allowing ultraviolet radiation to reach the ground much more intensely than was ordinarily the case (Farman et al., 1985). Subsequent scientific investigations soon led to what is now the most widely accepted explanation of what was happening. Chlorine compounds derived mostly from chlorinated fluorocarbon gases (CFCs), mass-produced by industrial societies for a variety of purposes, reacted in the stratospheric clouds over Antarctica during the cold, dark, winter months to produce forms of chlorine that rapidly deplete stratospheric ozone when the first rays of the Antarctic spring sunlight arrive (Solomon, 1990). Massive destruction of ozone followed very quickly, until natural circulation patterns replenished the supply and closed what came to be known as "the ozone hole." Human activities in distant areas of the planet had brought a sudden and potentially devastating change to the Antarctic and its ecosystems, a change that did not bode well for the ozone layer in other parts of the planet (Stolarski, 1988).

To understand this event and the political controversies that followed in its wake, one has to reach back through almost a century's worth of history, long before CFCs existed. Until almost the end of the nineteenth century, refrigeration was a limited technology, based almost entirely on natural sources of supply. Urban Americans who could afford to drink chilled beverages relied on metropolitan ice markets, which cut ice from local ponds in the winter and stored it in warehouses for use during the warm months of the year. Breweries and restaurants were the heaviest users of this stored winter ice, which was sometimes shipped hundreds of miles to provide refrigeration. Boston ice merchants, for instance, were regularly delivering ice to consumers in Charleston, South Carolina, and even the Caribbean by the fourth decade of the nineteenth century (Hall, 1888; Cummings, 1949; Lawrence, 1965).

   Given the expense and difficulty of obtaining this stored winter ice, food preservation was accomplished largely with chemical additives, the most common being ordinary table salt: sodium chloride. In the United States, pork was the most popular form of preserved meat because of the ease with which its decay could be arrested by salt. Beef was much less popular in preserved form, so those who ate it preferred to purchase it freshly slaughtered from local butchers. Then, in the 1870s, meatpackers began experimenting with ice-refrigerated railroad cars that could deliver dressed beef, slaughtered and chilled in Chicago, to consumers hundreds of miles away. Dressed beef, which was cheaper than fresh beef for a variety of reasons, soon took the country by storm, driving many wholesale butchers out of business and giving the Chicago packing companies immense economic power. The packers initially relied on complicated ice storage and delivery networks, cutting and storing millions of tons of winter ice along the railroad routes that delivered beef from Chicago to urban customers throughout the East. Their investment in ice storage technology contributed to dramatic shifts in the American food supply and was soon affecting foods other than meat. Fruits and vegetables from California and Florida and dairy products from metropolitan hinterlands throughout the East, were among the most important to benefit from the new ice delivery system (Cronon, 1991; Yeager, 1981; Kujovich, 1970; Giedion, 1948; Clemen, 1923; Swift and Van Vlissingen, 1927; Neyhart, 1952; Unfer, 1951; Fowler, 1952).

   But natural ice was unreliable: two warm winters in 1888-1889 and 1889-1890 brought partial failures of the ice crop that encouraged the packers to turn to a more reliable form of refrigeration. Although the principle of mechanical refrigeration, in which compressed gas was made to expand rapidly and so lower temperatures, had been known since the middle of the eighteenth century, its first application on a large commercial scale was not found until the second half of the nineteenth century (Anderson, 1953). Urban brewers, especially in the warm climates of the South, were the first to make wide use of it. As the meatpackers sought to solve their problems with erratic winter ice supply, they too adopted mechanical refrigeration on a large scale after 1890. By the first quarter of the twentieth century, the delivery of perishable foods throughout the United States—and international food shipments as well—had come to depend on mechanical refrigeration. By drastically lowering the rate at which food decayed and hence making perishable crops available to consumers through much of the year, refrigeration changed the whole nature of the American diet.

The most widespread early refrigeration technology depended on compressed ammonia gas, which easily produced desired drops in temperature for effective food storage. But ammonia (like other refrigerant gases such as sulfur dioxide and methyl chloride) had serious problems. For maximum efficiency, it had to attain high pressures before being released, which increased the likelihood that the compression equipment might fail. Accidental explosions were frequent, and the toxic nature of the gas caused a number of fatalities. Toxicity and the need for large expensive compressors kept mechanical refrigeration from making headway with retail customers, who represented an immense potential demand. That is why Thomas Midgely Jr.'s 1931 invention of Freon 12 represented a revolution for the refrigeration industry. Midgely, working at the request of the General Motors Frigidaire division, developed the new chlorinated fluorocarbon as the perfect alternative to all other refrigerant gases then on the market.

Nonflammable, nonexplosive, noncorrosive, and nontoxic, the various forms of Freon gas seemed the perfect technical solution to a host of environmental and safety problems. They also required less pressure to produce the desired cooling effect, so compressors could be smaller and less expensive. Freon soon came to dominate the market for refrigeration and opened up new retail markets because of its diminished capital requirements. Previously, consumers had bought their refrigerated food at the store just before eating it, since efficient and reliable household refrigeration was not generally available. Now American households could own their own refrigerators, making it possible for the food industry to shift much of its marketing apparatus toward selling chilled food in retail-sized packages. Frozen foods burst onto the American marketplace in the 1950s, as did fresh vegetables, dairy products, and other foods that are today accepted as ordinary parts of the national diet. Although European countries were slower to adopt these technologies, they too eventually followed suit.

No less importantly, the nontoxicity of Freon made it possible for refrigeration technology to be applied to the ambient cooling of buildings, so that air conditioning came to be an ever more important market for the gas. Air conditioning had been used in specialized industrial applications ever since Willis H. Carrier's use of the technique for a climate-controlled lithography plant in 1902. The introduction of Freon meant that air conditioning suddenly became much cheaper and safer in a way that allowed it to be applied to office buildings and finally to residences as well. Air conditioning played a key role in the years following World War II in promoting urban growth in the region known as the Sun Belt, as well as in tropical areas around the globe. From Florida to Texas to southern California, the massive influx of new residents depended in no small measure on the ability of buildings to protect their occupants from summer heat. Air conditioning became a fact of life in such places, so much so that it is hard to imagine urban life in the Sun Belt without it. Its significance can be captured by two phenomena of striking environmental significance: the shift in the seasonal consumption of electricity from peak load during the winter months (when energy consumption for lighting and space heating had always traditionally been at its highest) to peak load during the summer; and the steeply upward slope in the production and consumption of chlorinated fluorocarbons. The upward trend in CFC production was also aided by the development of still other uses for CFCs: as nontoxic propellants in aerosol sprays and later, in the 1960s and 1970s, as solvents in the manufacture of integrated circuits.

CFCs are very stable gases: that is in fact one of the properties that made them seem so benign when measured by their toxicity and immediate environmental effects. But the very stability that made CFCs so attractive for so many applications proved finally to be their greatest hazard. Once released into the environment— and the proliferation of refrigerators, freezers, and air conditioners meant that Freon escaped at an ever increasing rate—CFCs began to permeate the atmosphere, eventually reaching its upper regions. There they encountered the ozone layer, the thin belt of unstable tripartite oxygen molecules that filters out much of the sun's ultraviolet radiation and protects living organisms on the surface of the planet from the effects of that radiation. In the presence of sunlight, CFC molecules became chemical agents capable of destroying many times their number of ozone molecules. This effect was first hypothesized in 1974 by the chemists Mario Molina and Sherwood Rowland of the University of California at Irvine, writing in the wake of the controversy over supersonic transport aircraft and with recent knowledge, developed through new detection technology, that CFCs were present in the atmosphere (Molina and Rowland, 1974). Their hypothesis was controversial but convincing enough to produce action by the United States and eight other countries to ban the use of CFCs in aerosol sprays in the late 1970s (unquestionably the most marginal of their uses). Significantly, the suggestion that CFCs might possibly be damaging to the ozone layer did not have much effect on uses that were much more central to the industrial economy: food refrigeration, ambient air conditioning, and electronic manufacturing solvents. (The knowledge that CFCs account for a significant proportion of the human contribution to the greenhouse effect—about 25 percent by the mid-1980s—also did not have much effect.)

Not much effect, that is, until 1985 and the discovery of the ozone hole over Antarctica. Within two years' time, the scientific community agreed that CFCs were the most likely culprit; officials at DuPont, which produced 25 percent of the world's CFCs, declared the company's intent to phase out CFC production over the next decade and a half; and an international protocol was signed at Montreal, in which signatory countries declared their intention to cut CFC production and consumption in half by the end of the century (Benedick, 1989a, b; U.S. Office of Technology Assessment, 1988; Haas, 1989).

The lessons of this story about CFCs and the ozone hole are several. On the positive side, the rapid response of the scientific, industrial, and policy-making communities to the discovery of the ozone hole over Antarctica is reassuring proof that international agreements in response to global change are in fact possible. That the Montreal Protocol and the later, even stronger London amendments to it could be signed even in the absence of environmentally benign alternatives to the CFCs suggests people's perception of how serious and urgent the problem had become, but also their faith—encouraged by DuPont's actions—that alternatives would in fact be available by the time the agreement's deadline fell due. Indeed, the Montreal Protocol is a paradigmatic case of a quick technical fix, in which people respond to the environmental problems of a particular substance by finding (or hoping to find) a technology that can be used for exactly the same purposes without requiring any fundamental change in human economies or societies.

And that suggests some of the less reassuring lessons of this story. Refrigeration and air conditioning have today become so embedded in the American way of life, and in the ways of life of many people the world over, that it is hard to imagine modern food supplies and urban life styles without them. The very form of the post–World War II city, with its tall office buildings, fixed windows, and energy-intensive controlled climate systems, presumes a significant commitment to refrigeration and cooling. Almost no one has responded to the ozone hole by suggesting a retreat from these fundamental technologies of modern life: almost everyone assumes that existing technologies can be sustained more or less unaltered by introducing some other gas as an alternative to Thomas Midgley's 1931 invention. A quick technical fix may well be all that is needed, in which case the refrigeration-intensive (and energy-intensive and greenhouse gas-intensive) food and architectural systems of the twentieth century First World will continue to proliferate around the planet, with countries of the tropics presumably adopting them with even greater reason and greater intensity than those living in temperate regions.

Of course, such a chain of events might well accelerate global climate change. The invention of CFCs started a process that led to building practices and patterns of human settlement with two unexpected and long-term effects on the global environment: a built-in demand for CFCs and a built-in demand for energy, not only for space cooling but also for transportation to and between the new dispersed, warm-climate population centers. A quick fix for the effects of CFCs on the ozone layer might encourage the spread of the American pattern of energy-intensive settlement. A possible result is more rapid growth of greenhouse gas emissions than would otherwise be the case.

The encouraging policy success at Montreal in 1987 was dramatic, but may have depended on special circumstances: there were only about two dozen CFC producers worldwide, and reductions threatened few of the existing infrastructures that had developed over the previous century and a half. For that reason, the signing of the Montreal Protocol is a risky predictor of how other international negotiations may turn out when the response to global change seems to require greater alterations in historical practice, when there are many millions of responsible actors, or when the costs and benefits of change are less evenly distributed around the planet.

There is one final lesson of the CFC story that is most ironic of all. We would do well to remember that chlorinated fluorocarbons were themselves a response to serious environmental problems. They reduced the occupational hazard of compressor explosions, they all but ended toxic pollution (and deaths) from refrigerant gases, and they dramatically increased the variety and safety of the human food supply. For 50 years, they seemed a perfect example of a benign technical solution to environmental and engineering problems, with no negative side effects of any kind. We now understand that the very quality that made them seem so safe—their stability—means that they will continue to destroy ozone molecules far into the future even if we were to end their production and use at this instant.

The history of CFCs demonstrates, above all else, that human activities can have quite unexpected long-term effects on the environment. CFCs, initially developed to support a limited set of end uses in the refrigeration industry, have changed not only that industry but also significant aspects of human civilization. As a result, they have made major contributions both to stratospheric ozone depletion and to global climate change. Moreover, because CFCs have contributed to social changes that are built into national building stocks, transportation systems, and even political structures (congressional representation from the Sun Belt, for example), the indirect effects of CFCs on climate change may be very difficult to reverse, even if substitutes are found that do not harm the ozone layer. Dependence on refrigeration has created social pressures to resolve the ozone problem by technical means, a strategy that could have paradoxical results: the solution to the ozone problem could accelerate social processes that cause climate change. The CFC story demonstrates the tremendous difficulty of understanding the environmental effects of technological change. It suggests that connections need to be traced through a greater variety of technological and social systems and over longer periods of time than usually covered in social scientific studies. We return to these difficult long-term scientific challenges in Chapter 5. The CFC story also drives home the point that we cannot anticipate all the environmental or social effects of our own activities, suggesting that the best policies are those designed with considerable robustness to unintended consequences.

COAL COMBUSTION IN CHINA

Fossil fuel consumption accounts for over half the human contribution to the greenhouse effect, chiefly through the emission of carbon dioxide. Although the People's Republic of China is only the world's third-largest producer of carbon dioxide (after the United States and the Soviet Union), it is increasing its rate of production faster than any other country (750 million metric tons more in 1988 than in 1980—National Academy of Sciences, 1991a). Three-quarters of the Chinese emissions come from burning coal. The rapid increase in Chinese coal consumption—from 62 million tons (Mt) in 1952 to 812 Mt in 1985—can be traced to industrialization, electrification, and population growth (Xi et al., 1989). The trend seems likely to continue over the next several decades because China is in an energy-dependent phase of development and has few alternatives to coal. China has the world's third largest coal reserves, after the Soviet Union and the United States, but is very limited in reserves of other fossil fuels (Xi et al., 1989) and lacks the capital for major investments in nuclear power or development of its large, but inconveniently located, hydroelectric potential.

Causes of Present Coal Burning

A simple way to analyze energy use in China is to use the accounting equation:

$$E = P \times GNP/P \times E/GNP,$$

where E is energy consumption, P is population, and GNP is gross national product. Thus, energy use is the product of population, per capita economic output, and energy intensity—that is, energy use per unit of output. Chinese energy use in 1987 was 435 percent of what it was in 1965, while population was 147 percent, GNP per capita 305 percent, and GNP 97 percent of 1965 levels:

$$4.35 \ E_{1965} = 1.47 \ P_{1965} \times 3.05 \ GNP/P_{1965} \times 0.97 \ E/GNP_{1965}$$

(data from World Bank, 1989:Tables 1 and 5). This analysis suggests that roughly two-thirds of the rapid increase in Chinese energy use was a result of economic development, and the rest was due to population increase. But a closer look at the relationship of energy use and GNP gives a different picture—one that puts much more emphasis on technology and its social control.

China's energy use is a story not only of economic development, but also of persistently intensive energy use. China's economy is far more energy-intensive than that of most other countries or, put another way, China gets much less economic output from each unit of energy. Tables 3-6 and 3-7 show that China's economy may be the most energy-intensive in the world. In terms of CO_2 emissions per unit of economic output, China is by far the world leader (National Academy of Sciences, 1991a).

The few available analyses of energy use in China suggest that its energy intensity has two main sources: industrialization and inefficiency. Industry is more energy-intensive than other productive sectors, and China devotes a greater proportion of its recorded energy consumption to industry and is more dependent on coal in that sector, than most other countries (see Table 3-8). This pattern may be traceable to a Stalinist development policy that favors heavy industry on ideological grounds. The government, which determines production by directive rather than allowing it to respond to demand, is said to continue to command steel production, despite huge surpluses (Smil, 1988, and personal communication).

TABLE 3-6 Energy Intensities in Selected Countries and Groups of Countries, 1987

Country	Energy Intensity[a]	Productivity[b]
China	1.81	76
India	0.69	199
40 other low-income countries	0.41	335
53 middle-income countries	0.60	229
25 high-income countries	0.34	404

[a]Kilograms of oil equivalent per U.S. dollar of GNP.
[b]U.S. dollars of GNP per barrel of oil equivalent (1 barrel = 137.2 kg).
Source: Calculated from data in World Bank (1989).

TABLE 3-7 The Most Energy-intensive Economies in the World, 1987

Country	Energy Intensity[a]	Productivity[b]
China	1.81	$ 76
Poland	1.75	78
Yemen, People's Democ. Republic	1.68	82
Zambia	1.52	90
Hungary	1.37	100
South Africa	1.30	106
Trinidad and Tobago	1.23	112
Jamaica	0.91	151

Note: Complete data not available for Afghanistan, Albania, Angola, Bhutan, Bulgaria, Burkina Faso, Burma, Chad, Cuba, Czechoslovakia, German Democratic Republic, Guinea, Iran, Iraq, Ivory Coast, Kampuchea, Korea (Democratic People's Republic), Mongolia, Namibia, Romania, U.S.S.R., Vietnam, and countries with less than 1 million population.
[a]Kilograms of oil equivalent per U.S. dollar of GNP.
[b]U.S. dollars of GNP per barrel of oil equivalent (1 barrel = 137.2 kg).
Source: Calculated from data in World Bank (1989).

The main reason for energy intensity, however, appears to be an inefficiency that has several contributing causes:

1. *Inefficient End Uses* Coal burning in China is typically done in small, old units owned by households or small enterprises—characteristics that spell inefficiency. In the United States, 85 percent of coal is burned to generate electric power, at an average efficiency of 36 percent. By contrast, 22 percent of Chinese coal is converted to electric power, with an overall efficiency of only 29-31 percent (Kinzelbach, 1989; Xi et al., 1989). The bulk of Chinese coal is burned at still lower efficiencies, in industry (46 percent of 1985 coal use) and for commercial and residential heating (26 percent). Residential coal stoves often have only 10-18 percent efficiency (Xi et al., 1989). Adoption of more efficient furnaces and replacement of coal-fired space heating with combined heat and power installations proceed slowly for lack of capital.

2. *Price Structure* Policy sets coal prices for the state-owned mines artificially low, below the cost of production. Although the industry operates at a loss (Xi et al., 1989), the government is said to be reluctant to raise prices for fear of inflation and urban unrest. Many analysts see price as the key source of continuing

TABLE 3-8 Percentage of Commercial Energy Used for Industrial Purposes and Percentage of Industrial Energy Supplied by Coal in Selected Countries and Groups of Countries

Country/Group (Date)	Industrial Energy Use (%)	Direct Coal Use (% of Industrial Energy)
China (1980)	63	67
India (1984)	53	72
Brazil (1983)	45	35
Indonesia (1984)	44	1
Korea (south) (1985)	41	34
U.S.S.R. (1987)	33	28
Eastern Europe (1983)[a]	52	19
OECD members (1985)	36	26

[a]Bulgaria, Czechoslovakia, German Democratic Republic, Hungary, Poland, and Romania.

Source: Calculated from World Resources Institute and International Institute for Environment and Development (1988:Table 7.4).

inefficiency, in that efforts to improve efficiency in either mining or consumption look uneconomic with current prices.

3. *The Command Economy* The practice of government-dictated production, combined with the price structure, allows highly inefficient enterprises to continue operating despite financial losses. Enterprises that could compete by using energy more efficiently do not have incentives to do so. Moreover, the system of production quotas encourages the shipment of uncleaned, unsorted coal with an energy value of 30 percent less than actual tonnage (Smil, 1988). Such coal fulfills quotas easily, inflating production statistics by over 100 Mt per year, but it strains the Chinese railroads, 40 percent of whose cars are devoted to moving coal; wastes fuel in transport; and results in substantial emissions of unburned particulates when the coal is used.

Table 3-7 offers a rough guide to the amount of inefficiency a command economy can produce. Although data are available only for a few such economies, among these are four of the five least energy-productive economies in the world. The other large-population, low-income countries of the world, India, Indonesia, Nigeria, Bangladesh, and Pakistan, get 2.5 to 6 times as much production as China out of each unit of energy they use (data from World Bank, 1989). Although China cannot be expected to increase its energy productivity 2.5 times to India's level—the ample availability of low-cost coal in China gives it less incentive to economize on energy—it seems to have room for huge improvements in efficiency.

Determinants of Future Coal Burning

The future of global climate change depends very much on how energy-intensive future Chinese development will be. Between 1965 and 1987, Chinese coal use—and CO_2 emissions—increased at the same rate as total economic output. If both continue to increase at the recent historic rate of 4 percent per year, the Chinese contribution to global CO_2 emissions will quadruple in less than 40 years and surpass that of the United States, presuming that the latter also follows recent trends. However, if future economic growth can be less energy-dependent than past growth has been, the picture would be quite different (data from World Bank, 1989; Fulkerson et al., 1989).

What determines whether economic growth will or will not increase CO_2 emissions? Historical data show that successful economic development in Western countries has been marked by a period of rapid industrialization and consumption highly dependent on increased use of energy from fossil fuels, followed by a period in which economic growth becomes less energy intensive. Economic growth can proceed with decreasing energy intensity because of shifts of production from industrial to service sectors and adoption of more energy-efficient processes and technologies (World Resources Institute and International Institute for Environment and Development, 1988:114); energy use per capita, however, has continued to increase in these countries (World Bank, 1989:173).

The future of China's energy use can be analyzed in the terms of the accounting equation: population growth, economic development, and changes in energy intensity or productivity. A fourth factor—shifts from fossil fuels to other energy sources—is unlikely to have much influence in China for several decades unless there is a major international effort to promote such shifts.

Population growth, barring wars, epidemics, and the like, is easier than the other variables to forecast, because it is driven mainly by the current age distribution and slowly changing fertility trends. Beyond a few decades, though, uncertainties increase: desired family size may change, as may population policy, which has recently been holding family sizes below the levels parents seem to prefer. Forecasts for the year 2025 give a range of 1.4 to 1.6 billion for China's population (29 to 51 percent above the 1985 level) (United Nations, 1989).

Economic growth and energy intensity are closely interrelated and very difficult to forecast. The Chinese national growth plan calls for quadrupling GNP from 1980 to 2000, but for coal use to increase only 2.3 times (Smil, 1988; Xi et al., 1989). These forecasts call for the elasticity of energy/GNP to decrease from the 0.97 of the 1965-1987 period to about 0.6, a change that would save more fossil fuel in 2000 than China used in 1985, if the quadrupling of GNP is achieved.

There are tremendous uncertainties in predicting whether these goals can be met, even though Chinese energy use is certainly inefficient enough to allow this much technological improvement. Some observers believe the goals can be met only after continued economic liberalization, including price reform and market incentives, and political reforms that would overhaul wasteful management practices and attract needed foreign technology, expertise, and capital (e.g., Smil, 1988). The probability of such policy changes is notoriously uncertain, as the political events of 1989 in China and Eastern Europe attest. And given the current level of knowledge about the functioning of command economies, even if policy changes were known in advance, the success of their implementation, and therefore their precise effects on energy productivity, would be hard to predict.

No one knows how the Chinese will use the fruits of future economic development. If they make major investments in energy productivity—for instance, modernizing the coal industry, using electricity to replace inefficient coal burning, and developing the service sector of the economy—much can be done to mitigate CO_2 emissions. But other directions of investment, focusing on new manufacturing and expanded energy services such as refrigeration and personal transportation, would be much more energy intensive. If China makes a major shift toward market incentives, the decentralization of choice will promote efficiency in production, but it might also encourage energy-intensive consumption, as individuals gain disposable income. The net effect on energy intensity is still unknown.

Another important unknown is whether government policies will emphasize energy efficiency and the global environment. China already has policies to reduce coal use, but not in order to improve energy efficiency. The priorities are urban air pollution, freeing rail cars for noncoal cargo, and reduction of sulfur oxide emissions. These priorities encourage some energy-productive investments, such as combined heat and power plants that capture waste heat to warm buildings. But other important energy-productive investments do not fit these priorities. The future thrust of Chinese environmental policy depends, of course, on politics. Current environmental policies have been set from the top down, influenced by the exposure of traveling Chinese officials to the environmental concerns of foreign scientists, international organizations, and investors (Ross, 1987). If China turns inward to resist democratization, global concerns about energy efficiency may not influence Chinese policy for a long time. If environmental politics in China decentralizes and democratizes, an opening will appear for local environmental movements, which have been prevented from forming horizontal linkages in the past (Ross, 1987). Freedom for Chinese environmentalists, however, might lead to pressure for local changes, rather than for policies that improve energy efficiency nationally.

In sum, the Chinese contribution to global climate change depends on the interactions of technology with social factors, including population growth, economic development, policy, and ideology. Scientists know much about the technical changes that could mitigate China's greenhouse gas emissions, but they have relatively little quantitative understanding of the social factors that make possible, and interact with, technological change. Enough is known to identify some of the critical determinants of Chinese energy intensity, but not to quantify their effects or specify their interactions. That will require further research. For example, critical changes in policy, such as increased emphasis on market incentives and decentralized decision making, might greatly improve energy productivity. Studies of transitions to increased market control in other command economies might provide valuable knowledge for projecting the likely effects of such policy changes on energy efficiency in China. The future of Chinese energy demand also depends on changes in the structure of the Chinese economy and of consumer demand. Careful comparative studies of the social determinants of energy intensity and changes in energy intensity at the level of nation-states are critical for understanding and projecting China's future contribution to the greenhouse effect.

FOREST CLEARING IN THE AMAZON BASIN

Clearing of tropical forests is generally considered to be the most important single cause of recent losses in the earth's biological diversity. It also accounts for about 15 percent of the effect of human greenhouse gas emissions. Clearing has been very extensive in recent years, and the disturbances are not readily reversible, as deforestation by indigenous slash-and-burn techniques had previously been (Conklin, 1954; Nye and Greenland, 1966; Sanchez et al., 1982). The damage is now so extensive and severe as to preclude regeneration to original cover without special measures that are only now being developed (Uhl et al., 1989).

The most widely used definition of biological diversity includes three levels: genetic, species, and ecosystem diversity (Norse et al., 1986; U.S. Office of Technology Assessment, 1987). Deforestation reduces diversity at all three levels. Genetic diversity, or the diversity of genes within a species, provides the raw material for evolution, as it allows some individuals of a species to survive environmental changes that prevent other individuals from living or reproducing. Species diversity, which refers to the many million species now estimated to exist on earth, is richest in the tropical forests, particularly in the Amazon Basin (Erwin, 1988). Many of the Amazonian species are closely tied to particular forest ecosystems and tree species, so that they are very narrowly distributed and especially vulnerable to extinction by regional or even local forest clearing. Ecosystem diversity, that is, the existence of distinctive communities of species in different physical situations based on factors such as soil types or height above the river channel (Prance, 1979), is also great in the Amazon Basin, even between physical situations that look identical to the untrained eye.

Amazonian deforestation threatens these forms of biological diversity in many ways. Elimination of forests destroys the habitat of many species that are closely tied to particular trees or ecosystem types. Species whose habitats are not totally destroyed may become extinct when an insufficient number are left in the remaining habitat, or remaining patches do not contain the resources they need (such as nest sites or food from a particular tree necessary to sustain the species). Species may be eliminated because of ecosystem simplification, as when removal of a single species eliminates the many species dependent for their existence on the local population of that species,[1] or when cutting eliminates the cool, moist, windless microclimates of the forest interior that many species require. Diversity is vulnerable to drying of the regional climate, because evapotranspiration from the forest generates about half the rainfall in the Amazon Basin (Salati and Vose, 1984).[2] Deforestation can damage biological diversity by contributing to both global and regional climate change, especially if the result is a drier climate in the Amazon Basin. Road building, in addition to destroying the forest, increases access to it and facilitates further deforestation. Deforestation favors species that occur only in highly disturbed areas, such as weeds, mosquitoes, and cattle, and that spread disease, compete with native organisms, and change the soil structure (Denevan, 1981). Finally, much deforestation is a by-product of industries such as mining, which not only destroys the forest at the industrial site, but may also use large numbers of trees for fuel.

Deforestation reduces biological diversity in several ways. In general, the species hardest hit are likely to be the ones with large area requirements, narrow ranges, or value to humans for food, medicine, or timber, yet the entire taxonomic spectrum may suffer major losses.[3] Some threatened species may be important to the region's economy and culture, some are used far beyond the Amazon Basin, and some have potential value to humans that is not yet known. The threatened ecosystems provide regionally important services, such as creating soils, moderating temperatures, reducing soil erosion, cleaning the air and water, and preventing flood and drought (Smith, 1982). The net effect on humans is impossible to estimate in advance but, whatever its size, it is likely to be irreversible.

Causes of Deforestation

Amazonian forest land is cleared for many purposes. Logging is a major industry, with four of the six states in the Brazilian Amazon Basin depending on wood products for more than 25 percent of their industrial output (Browder, 1988). Other industries destroy forest both as an integral part of the manufacturing process and as a by-product. For example, 610,000 hectares (ha) of forest per year are used to produce charcoal for iron smelting in the Gran Carajas region of Brazil (Treece, 1989). Damming rivers to generate electricity for aluminum refining and for urban power inundates huge areas because of the low relief of the land. But the largest single source of Amazonian deforestation and the focus of this discussion is cattle raising, which now covers an estimated 72 percent of the cleared area (Browder, 1988).

The transformation of forest into inferior, rapidly degrading pasture was not inevitable. It was strongly influenced by national policies and supported by international development agencies, which encouraged migration and land clearing through land-titling arrangements, provided a publicly financed infrastructure of roads, and established credit and tax incentives to benefit ranching. Given these institutional conditions and the presence of abundant, accessible, and relatively cheap land in the Amazon, individual actors made rational economic choices that furthered their own best interests and helped create a system with its own economic and social momentum that continues deforestation even after state incentives have been removed.

Road Building With support from the World Bank, the Interamerican Development Bank, and other international lending institutions, the Brazilian government improved and paved major north-south (Belem-Brasilia) and east-west (Cuiaba-Porto Velho) highways, hoping to tap the wealth of the Amazon, make minerals and timber accessible, and promote agricultural enterprises (Fearnside, 1989). If, as the planners intended, settlers had migrated from the poor, drought-stricken northeast to settle along the trans-Amazon highway, they might have developed the area intensively, with permanent, smallholder farming and agroforestry, and limited deforestation. But mass migration did not occur in the northeast, and much of the area was abandoned to pasture

(Browder, 1988; Moran, 1976, 1990). The opening of new lands and the relative absence of people favored extensive development, such as ranching, over intensive development.

Land Tenure Rights   For centuries, it has been the legal practice to grant rights of possession to whoever deforests a piece of Brazilian land. Rights of ownership soon follow (Fearnside, 1989). Squatters on public land can gain the rights to 100 ha by living on it and using it, but 100 ha is not sufficient for ranching. Ranchers often buy up the lots of failed farmers, and in 1974 it became possible for a company to acquire a tract of up to 66,000 ha (Smith, 1982). Large individual and corporate ranchers can build their own access roads and lay claim to extensive plots far from major highways. By the time roads are constructed, most state land in the Amazon is already claimed (Binswanger, 1989). Brazilian land laws encourage both extensive holdings and extensive use. For instance, the 1988 constitution provides that land "in effective use," that is, cleared, cannot be expropriated for the purpose of agrarian reform (Hecht, 1989b).

Speculation   Land holding has been a useful hedge against Brazil's galloping inflation and an excellent speculative investment. Mahar reports that a farm laborer can "net the equivalent of $9,000 from clearing 14 ha of forest, planting pasture, and a few crops for a few years, and selling the 'improvements' to a new settler" (1988:38). This is four times what the laborer could hope to earn from ordinary farm work. The largest speculative gains accrue to large investors with good connections in government and the courts because the value of land is greatly influenced by "institutional factors such as validity of title, [and] access to credits" (Hecht, 1989b:229).

Financial Incentives from Government   To encourage development in the Amazon, the Brazilian government made rural credit available to those with a land title or a certificate of occupancy at low, indeed at negative, interest rates. The credits were so attractive that money flowed from the nonagricultural sector into extensive ranching (Binswanger, 1989). Small farms were not taxed on land, large ones could reduce their already low taxes by converting forest to pasture or crops (Binswanger, 1989), and corporations could deduct up to 75 percent of the cost of approved development projects in the Amazon from their federal tax liability (Browder, 1988). Corporations could also write off losses on Amazon projects against taxable income earned elsewhere (Browder, 1988). These incentives favored extensive enterprises and encourage livestock production even when returns from beef alone did not pay the cost of production. Fiscal incentives for livestock raising have largely dried up since 1985, but the cattle population has continued to grow at an annual rate of 8 percent (Schneider, 1990), suggesting that profit can now be made without subsidies, partly from the appreciation of land values (Binswanger, 1989).

Livestock and Crop Economics   The strategy that is generally most immediately profitable when land is plentiful and labor scarce is one of extensive and often transitory use. An example is shifting cultivation, the predominant indigenous strategy of land use. Fire removes cut brush and trees, and there is no need to turn the soil, weed frequently, irrigate, drain, or terrace. Beef production demands even less work per unit output and, with the help of modern technology and fossil-fuel energy for clearing forests, can be much more extensive than shifting cultivation. Fattening cattle on grass requires little labor or expenditure on fencing and corrals, and no weeding. Ranchers can take advantage of the highly productive first years after forest clearance to overstock the range and increase short-term profit. Cattle projects supported by the Brazilian development agency, SUDAM, operated with as few as one employee per 400 ha (Denevan, 1981). Such ranches, established with government subsidies, are now able to survive without them by marketing more timber from the land, selling beef to recent migrants to the new urban centers in the region, walking their cattle to market, and using new and better-adapted grass species and selectively bred cattle (Schneider, 1990).

Ideology, Politics, and Economics of Development   Throughout much of the 1960s and 1970s, the Brazilian government with support from international financial institutions pursued a strategy of large-scale, capital-intensive development projects. These often meant monocropping, relatively low labor inputs, mechanization, and the maximization of short-term financial returns. The strategy, elaborated in textbooks on development (e.g., Rostow, 1960), was based on shared premises about the essential goodness of economic growth and made deforestation for lumbering and extensive cattle raising the most profitable activity on the Brazilian frontier (Partridge, 1984). The international debt incurred in part to promote such development increased demands for rapid returns, high profits, and the production of exports to pay the interest charges (Hildyard, 1990). Recently, disappointing economic returns, declining international aid, and an awareness of rapid ecological deterioration are becoming associated with changing priorities, and analysts in the World Bank and elsewhere are becoming critical of the old development philosophy (Binswanger, 1989; Mahar, 1988; Schneider, 1990).

The Role of Population Growth It is easy to see Brazil's average population growth of 2.8 percent in the 1970s as the source of land hunger and migration, raising the Amazon population by 6.3 percent annually (Browder, 1988). However, the period witnessed stronger movements of population from the already settled hinterland to cities, combined with considerable natural increase in urban areas. The decline in rural population density is reflected in the phrase, "*Quando chega o boi, o homen sai,*" (When the cattle arrive, the men leave) (Browder, 1988:254). The extensive clearing of forest on the frontier reflects population pressure and food needs *outside* the local region, combined with a lack of population pressure locally (Denevan, 1981).

Alternative Futures for Amazonia

The Amazonian case illustrates the difference between intensive and extensive patterns of land use in tropical forests. Table 3-9 provides a summary representation of the extremes of these patterns, presented as ideal types (actual land use almost always has features of both types). The Amazonian forest has long been inhabited by peoples that used a mixture of these strategies to support their economies. Indigenous groups combined relatively extensive strategies, such as temporary or shifting cultivation followed by natural forest regeneration and hunting and gathering of dispersed game, fish, and wild food plants, with more intensive farming of alluvial riverine and other soils of high, renewable fertility. More recently, both native American (Posey, 1989; Prance, 1989) and immigrant populations such as the rubber tappers have maintained the forest by a mixed-management strategy that mimics rather than replaces the biologically diverse natural environment (Browder, 1989).

The modern forms of land use most implicated in deforestation—cattle ranching, crop agriculture, and logging and other industrial uses—are extensive and rapidly expansive, market and capital dependent, specialized in one or a few commodities, and mechanized or labor saving. Some observers point to modern strat-

TABLE 3-9 Land Use Type

	Extensive/Expanding	Intensive/Stable
Population	Low density, growing migratory	High density, permanent settlement
Agricultural Production	Low average yields, high variability, low crop diversity	High average yields, low variability, high diversity (cereals, tubers, vegetables, trees, livestock)
Size of Economic Units	Large, tendency to increase land area	Small, balanced fragmentation and accretion
Labor	Low total inputs, seasonally variable, unskilled, high returns per hour, often hired	High total inputs, steady inputs throughout year, skilled, low returns per hour, often household
Technology	Mechanical, energy imported, nonrenewable, capital intensive	Simple, often manual, energy local, renewable, labor intensive
Market Integration	High, output sold, inputs largely purchased, national and international commodity markets	Subsistence combined with cash production, not totally dependent on market prices, some purchased inputs
Land Tenure	Private, land values speculative but initially low, legal access politically determined	Private *and* common property rights, land values high, inheritance important, legal protections
Social Inequality	High, growing polarization, landlord elite and landless wage laborers	Moderate, stratified, smallholders significant, social mobility
Resource Dynamics	Extraction	Recycling
Response to Environmental Change	Vulnerability, boom/bust cycles	Resilience, buffering, flexibility
Environmental Impact	Degradation, decline in biodiversity, nutrient loss	Sustainability, fertility renewed, conservation

egies of mixed development as an alternative way of using the forest for human needs. They claim that intensive, stable agricultural land use with a mixture of crops and livestock can be combined with labor-intensive efforts to maintain soil quality by careful, thorough tillage, agroforestry, manuring, terracing, irrigation, and drainage. Thus they can provide high, reliable, sustainable production from smallholdings with high inputs of household labor and little capital or fossil fuel energy. These systems may also help preserve mature forest ecosystems from destruction by reducing development pressure on them (Anderson, 1990).

The potential for a future of less-extensive forest use in the Amazon Basin relates in part to land distribution. Inequality of land holdings in Brazil has increased greatly over the last few decades, with 70 percent of Brazilian farmers now landless and 81 percent of the farmland held by just 4.5 percent of the population (Hildyard, 1990). This pattern of increasing inequality also holds in the Amazon, making access to resources more difficult for subsistence farmers and hunters and gatherers and threatening indigenous land tenure systems based on communal rights (Chernela, n.d.; Poole, 1989). Larger landholdings bring more extensive use. In Pará state, for instance, small farms cultivate an average of 50 percent of their area, while farms of over 1,000 ha cultivate only 26 percent (Hecht, 1981). More intensive cultivation means that less forest must be displaced to meet human needs. Moreover, stable smallholders have an incentive to economize on land and keep it productive, so that land degradation can be slower with more intensive use. Thus, the current pattern of extensive development, by displacing indigenous peoples and small-scale extractors, has removed a brake on deforestation and threatens a store of valuable knowledge about the intensive management of forest species for human consumption.

There are barriers to a transition to a mixed-development strategy in the Amazon. One is the social change resulting from the current extensive strategy. Another is the politics of change. With rural poverty increasing and a political choice between dividing up large landholdings and encouraging the landless to colonize unclaimed or "unused" frontier lands, migration and resettlement policies are much the more palatable alternative (Macdonald, 1981). And finally, there are intrinsic social limits. Although portions of the local environment could support intensive land use like that of the wet rice/garden systems of south China and Java, the necessary local density of population with plentiful labor and nearby markets are not present (Moran 1987:75). Extensive, extractive land use with deforestation is likely to remain the most economically feasible and politically viable development strategy in the Amazon region because vast areas of cheap land are accessible and markets are distant.

In sum, the causes of Amazon deforestation lie partly in the same frontier conditions that have led to extensive land use in nineteenth century North America and elsewhere. In addition, development policy around the world has supported capital-intensive development of export monocultures. The unique institutional and political history of Brazil has helped determine the particular development pattern there, a pattern significantly different from that of tropical forest development in Zaire or Indonesia (Allen and Barnes, 1985; Brookfield et al., 1991; Lal et al., 1986). A key to the future of the forests lies in policy changes that could limit deforestation and extensive land use while increasing food production from existing agricultural areas. However, the social and economic changes brought about by Amazonian development have created barriers to making and implementing such policies.

EXPLAINING THE PROXIMATE CAUSES: SOCIAL DRIVING FORCES

The examples above illustrate how the proximate causes of global environmental change result from a complex of social, political, economic, technological, and cultural variables, sometimes referred to as *driving forces*. They also show that studies of driving forces and their relationships have been and can be done (National Research Council, 1990b; Turner, 1989). However, little of this research has been conducted on a global scale, for at least three important reasons: demand for such studies is a very recent phenomenon; relevant data at the global level are scarce; and social driving forces may vary greatly with time and place. Consequently, much additional work is needed to support valid global generalizations.

We distinguish five types of social variables known to affect the environmental systems implicated in global change: (1) population change, (2) economic growth, (3) technological change, (4) political-economic institutions, and (5) attitudes and beliefs.

Vocal arguments have been made for each of these as the exclusive, or the primary, human influence on global environmental change. In each instance, supportive evidence exists below the global level. Evidence at the global level, however, is generally insufficient either to demonstrate or dismiss claims that a particular variable causes global environmental change or is more important than some other variable.

We briefly outline the evidence supporting and qualifying claims that each class of variable is an independent influence on global environmental change. Our citations are not meant to be exhaustive, but rather to refer the reader to typical sources and critiques of claims about the importance of particular variables. For many of the authors cited, links between key explanatory variables and global environmental change are only implicit; in such instances, we draw out the implications for global environmental change. We also outline some of the key unanswered but researchable questions regarding these driving forces.

POPULATION GROWTH

Of all the possible driving forces of environmental change, none has such a rich history in Western thinking as population growth. Starting with Malthus, scholars have attempted to understand the effects of population growth on resource use, social and economic welfare, and most recently the environment. Few debates in the social sciences have been so heated or protracted as that around the impacts of population growth. Clearly, each person in a population makes some demand on the environment and the social system for the essentials of life—food, water, clothing, shelter, and so on. If all else is equal, the greater the number of people, the greater the demands placed on the environment for the provision of resources and the absorption of waste and pollutants. Stated thus, the matter is a truism. The source of controversy centers around more complex questions. Does all else remain equal in the face of population growth? Do simple increases in numbers account for most of the increase in environmental degradation in the modern world? Can population growth occur without major environmental damage? If not, is population growth a root cause of the degradation that follows, or merely an effect of more deeply underlying causes, such as changes in technology and social organization?

Ehrlich and others (Ehrlich, 1968; Ehrlich and Holdren, 1971, 1988; Ehrlich et al., 1977; Ehrlich and Ehrlich, 1990) hold that population growth is the primary force precipitating environmental degradation. They argue that the doubling of the world's population in about one generation accounts for a greater proportion of the stress placed on the global environment than has increased per capita consumption or inefficiencies in the production–

consumption process. They do not hold that other factors are unimportant in placing stress on the earth's resources and on the biosphere, only that population growth must be considered primary, because if all other factors could be made environmentally neutral, population growth of this magnitude would still spur resource stress and environmental degradation. Indeed, it is argued that once population has reached a level in excess of the earth's long-term capacity to sustain it, even stability and zero growth at that level will lead to future environmental degradation (Ehrlich and Ehrlich, 1990).

The critiques of this position are many. One strand of criticism argues that technological and socioeconomic factors are primary (e.g., Coale, 1970; Commoner, 1972; Harvey, 1974; Ridker, 1972a; Schnaiberg, 1980). Another criticism comes from those who argue that population, though it may be a driving force of change, is not necessarily a driving force of degradation (Boserup, 1981; Simon, 1981; Simon and Kahn, 1984). Rather, they view population growth as a driving force of improvement, which increases the capacity of society to transform the environment for the better, or as a reflection of society's success in improving the environment so as to support greater numbers. These critics offer evidence from long sweeps of history, such as the relationships between major sociotechnical changes in society and global increases in population (Deevey, 1960; Boserup, 1965). Others have suggested that these population increases are also associated with increasing global environmental change (Whitmore et al., 1991).

Since World War II, concern with rapid population growth has motivated the U.S. government, private foundations, and multilateral aid agencies to fund a substantial body of research on the causes of population growth. In addition to supporting individual studies, these bodies have devoted substantial resources to institutional development by subsidizing education, professional journals, and centers of excellence. The result has been impressive in building demography as a respected, interdisciplinary field within the social sciences, and in gaining knowledge of the causes of population growth. As we note in Chapter 7, this experience provides a useful model for advancing interdisciplinary social science research on global change.

Research on the causes of population growth provides some useful insight into the causes of global change and strategies to deal with them. For example, current fertility and mortality patterns suggest that world population will continue to increase well into the next century. But if fertility declines as fast throughout the developing world as it has in a few developing countries, this growth might be reduced by almost 2 billion people by the time that the population of the developing world would otherwise have reached 8 billion (World Bank, 1984). This research helps clarify how much growth is more or less inevitable because of the momentum built into the age structure of the world population.

Compared with research on the causes of population growth, very little research has been devoted to understanding its consequences for environmental quality. This is ironic, because it is concern with the consequences that motivates much support for research on the causes of growth. There is some research on the effects of population growth on economic growth and social welfare, though the topic is still subject to some controversy (much of this literature is summarized in National Research Council, 1986). Only a handful of empirical studies have examined the effects of population growth on the environment, and many of these are quite dated (e.g., Ridker, 1972b; Fisher and Potter, 1971). As a result, it is difficult to assess just how important population may be as a driving force. For example, in 1986 a National Research Council study committee composed of economists and demographers concluded that slower population growth might assist less-developed countries in developing policies and institutions to protect the environment, but could find little empirical work on the link between population growth and environmental degradation (National Research Council, 1986).

Research Needs

We believe an extensive research program is needed to explicate the environmental consequences of population growth and provide a sounder basis for deciding what actions may be appropriate in response. Such research should begin by acknowledging that the environmental consequences of population growth depend on other variables. For instance, a population increase of people with the standard of living and technological base of average North Americans in 1987 would use 35 times as much energy as an increase of the same number of people living at India's standards—and their respective effects on the global climate would be in roughly the same proportion. The critical questions for research, then, are about the conditions determining the environmental effect of a projected population increase at a particular place and time. What are the multipliers that represent the environmental impacts of a new person in a particular year and country? To what extent are multipliers such as annual income or annual distance traveled constant for a country, and to what extent are they contingent on other factors that may change over time, such as the manufacturing intensiveness or energy supply mix of the country's economy or the country's policies on income distribution or energy development?

ECONOMIC GROWTH

Global economic growth, defined as increases in the measured production of the world's goods and services, is likely to continue at a rapid rate well into the future. The human impulse to want more of the material things of life appears to be deep-seated, and the areas of the world in which people are most lacking in material goods are those with the greatest—and most rapidly increasing—population. Assuming United Nations and World Bank projections for world population to double to about 10 billion in about 50 years, with 90 percent or more of that growth occurring in the developing countries of Africa, Asia, and Latin America, and assuming that per capita income grows 2.5 percent and 1.5 percent annually in the developing and developed countries, respectively (a low projection, in both cases, by standards of the last several decades), global economic output would quadruple between 1990 and 2040.

Under these conditions the relative gap between per capita income in developing and developed countries would narrow, but the absolute gap would increase substantially. To the extent that per capita income aspirations in the developing countries are driven by comparison of their incomes with those in developed countries, aspirations for additional income growth in the developing countries may be even stronger in 50 years than they are now.

Increased income or economic activity as measured by such indicators as gross national product is not, of course, equivalent to increased well-being. There is considerable debate in the economic literature on how to measure welfare, focused on such questions as how to count things people value that are not traded in markets and whether expenditures for pollution control should be considered an addition or a subtraction from net welfare (e.g., Daly, 1986; Repetto et al., 1989). Although these questions are very important for analyzing human–environment interactions, most current analyses of the effects of economic growth and environmental quality are based on conventional definitions of economic activity.

Economic activity has long been a major source of environmental change and, for the first time in human history, economic activity is so extensive that it produces environmental change at the global level. The key issues concern the extent to which current and future economic activity will shape the proximate causes of global change.

The production and consumption of goods and services is bound by a fundamental natural law—the conservation of matter. Whatever goes into production and consumption must come out, either as useful goods and services or as residual waste materials. Since the conversion of inputs to useful outputs is never entire, it is fair to say economic activity inevitably stresses the environment by generating residual wastes.

Wastes must be disposed of somewhere in the environment. Economists note that disposal presents no important social problem if it is managed to reflect its true social costs and to be equitable in the sense that the costs are borne by those who generate the residuals. However, true social costs can be very difficult to determine, especially when wastes alter biogeochemical processes that are poorly understood. And when the wastes are released to the atmosphere, rivers, and oceans, it is difficult to ensure that those who generate the waste pay the costs. The problem of defining social cost and the separation of those who generate the costs of waste disposal from those who bear them are the keys to the waste-induced environmental problem (Kneese and Bower, 1979).

Economic growth also depletes the stock of nonrenewable natural resources such as coal, oil, natural gas, and metallic minerals and, in some cases, the stock of renewable resources as well, as when the rate of soil erosion exceeds the rate of restoration of soil and nutrients. Environmental degradation follows when extraction disturbs land or biota and when resource use generates wastes. Economic growth may also destroy aspects of the natural landscape, for example, pristine wilderness areas or vast geological features such as the Grand Canyon. Continued use of depletable resources will create economic pressure to develop renewable energy resources, expanded recycling, and substitute materials (see, e.g., Barnett and Morse, 1963; Smith, 1979; Simon and Kahn, 1984), but a quadrupling of the global economy over 50 years would result in continued resource depletion.

Depletion of nonrenewable resources need not threaten long-run economic growth if management of the resources takes adequate account of their future value and the likelihood of finding substitutes. This condition may be easier to meet than the conditions for managing wastes because it is much easier to establish clear, enforceable property rights in nonrenewable resources and because such property rights permit creation of markets that provide price signals of changing resource scarcity and incentives to take future as well as current resource values into account. Property rights are relatively easy to establish because, unlike in the atmosphere and the oceans, nonrenewable resources are localized, spatially well defined, and fixed in place.

But markets in nonrenewable resources are no panacea for the environmental effects of minerals extraction or fossil energy use. Current markets have no sure way to anticipate, and therefore reflect, the value future generations will put on the depleted resources. This is the issue of intergenerational equity in resource management, and there are strong arguments that markets cannot deal adequately with the issue (Sen, 1982; Weiss, 1988; MacLean, 1990). The values future generations will hold can only be guessed at, drawing on human experience so far. Given this uncertainty, most analysts advocate more cautious resource management than what current market signals indicate.

So economic growth necessarily stresses the environment directly by increasing quantities of wastes and indirectly by depleting resources. However, the relationship between economic growth and environmental stress is not fixed. The key analytic questions concern the conditions under which a given amount of present or future economic growth produces larger or smaller impacts on the environment.

Several conditions apply. It matters which pattern of goods and services is produced. An economy heavily weighted toward services appears to generate fewer wastes and less resource depletion per unit of output than one weighted toward manufactured goods. Experience so far indicates that consumption patterns shift toward services as per capita income rises, suggesting that the process of growth itself may induce less than proportional increases in environmental stress. It seems that past some point, consumers use their economic resources to purchase well-being that is decreasingly dependent on material goods (see Inglehart, 1990). If the historic pattern holds, future economic growth in the low-income developing countries will be materials and energy intensive for quite some time before a transition to a service economy sets in. But this projection is uncertain because of incomplete knowledge about the causes of that transition and the ways it might be altered by deliberate action.

Other shifts in economies can also change the relationship between economic growth and environmental quality. Per capita use of many materials has been declining in North America and western Europe for some time (Herman et al., 1989). Waste management based on recycling, redesign of production processes, and the treatment of the wastes of one process as raw materials for another can reduce the environmental impact of economic activity (e.g., Ayres, 1978; National Research Council, 1985; Haefele et al., 1986; U.S. Office of Technology Assessment, 1986; Friedlander, 1989). And an observed trend in the United States, in which the main source of pollution has shifted from production activities to consumption activities, has effects on the overall economy–environment relationship that are not yet clear (Ayres and Rod, 1986; Ayres, 1978).

The environmental effect of economic growth may also depend on forms of political organization. The comparison of emissions of CO_2 and pollutants in Eastern and Western Europe suggests that democratic countries may be able to deal more effectively with the effects of wastes than nondemocratic countries. When people who feel the effects, or become concerned about the effects on others, have ready access to political power, their concerns may possibly have more influence on policy. If this hypothesis is correct, then political trends toward democracy, such as in Eastern Europe, will tend to reduce the amount of degradation resulting from economic growth there.

National policies also help determine the environmental costs of economic growth. In many developing countries, policies have favored extensive use of "unused" resources and "underpopulated" land to increase national power and improve the welfare of their citizens. Countries such as the United States, Canada, Argentina, and Australia had such policies during rapid development phases, and other countries have followed the example. This model of development through frontier occupation and rapid creation of wealth required cheap food and raw materials from rural areas, an infrastructure of roads and transport to open up these areas, and huge infusions of capital for enterprises and settlement. An alternative development model generates increased production per unit of land by agricultural intensification rather than by extensive land uses such as shifting agriculture or ranching (Boserup, 1990; Turner et al., n.d.). Development of this kind can be carried out in a sustainable manner (Conway and Barbier, 1990; Subler and Uhl, 1990).

Research Needs

The effects of economic development on the proximate causes of global change appear to be contingent, among other things, on the structure of consumer demand, the population and resource base for agricultural development, forms of national political organization, and development policies. However, the nature of these contingent relationships, particularly the relationships between policy and the other variables, is not understood in detail. Research is critically needed on the ways consumer demand changes as income increases, the effects of national policies on patterns of production and consumer demand, the effects of agricultural intensity on economic growth and the environment, and the causes of shifts from more to less energy- and materials-intensive economies. These questions call for research both within and across the boundaries of disciplines and academic specialties.

Technological Change

Technological change affects the global environment in three ways. First, it leads to new ways to discover and exploit natural resources. Second, it changes the efficiency of production and consumption processes, altering the volume of resources required per unit of output produced, the effluents and wastes produced, and the relative costs and hence the supply of different goods and services. Third, different kinds of technology produce different environmental impacts from the same process (e.g., fossil-fuel and nuclear energy production have different effluents). Some technologies have surprising and serious secondary impacts, as the history of refrigeration illustrates (see also Brooks, 1986).

In one view, technological development tends to hasten resource depletion and increase pollutant emissions. In this view, technology as currently developed is a Faustian bargain, trading current gain against future survival (e.g., Commoner, 1970, 1972, 1977). Modern technology is seen as a much more significant contributor to environmental degradation than either population or economic growth. One reason is that modern technological innovation progresses much faster than knowledge about its damaging effects, both because the effects are intrinsically difficult to understand and because the powerful economic interests that benefit from new technologies influence research agendas to favor knowledge about the benefits over knowledge about the costs (Schnaiberg, 1980).

Three arguments are advanced to oppose or qualify the Faustian theme. In the first, technology's contribution to environmental change is deemed relatively unimportant (Ehrlich and Holdren, 1972). In the second, technological innovation and adoption are seen as induced by other forces, particularly demand from population (Boserup, 1981) or market forces (Ruttan, 1971) and therefore not a driving force. The third argument is that technological change is a net benefit to the environment because it can ameliorate environmental damage through more efficient resource use and the lessening of waste emissions (e.g., Simon and Kahn, 1984; also Ausubel et al., 1989; Gray, 1989; Ruttan, 1971).

These contradictory arguments, all plausible, can be weighed only by research that is specific (e.g., which technology, in which society, at what time) and that takes into account the other major social forces that cause or are affected by technological progress. For instance, technological progress is affected by the relative prices of energy, materials, and labor, with inventors and entrepreneurs having a built-in incentive to develop technologies that economize on the more expensive factors of production. As a result, technological development starting in countries with low-cost energy will be more energy intensive than technologies developed in countries in which energy is expensive and therefore more likely to have negative environmental effects. The effects of technology on the environments of poor countries may reflect the fact that much of the technological innovation adopted in poor countries originated in rich countries, which face different economic and environmental problems. National economic policies, as well as environmental and energy policies, can favor particular kinds of technological innovation and thus hasten or forestall environmental degradation. In the United States, debates about apportioning government energy research funds between nuclear, fossil, conservation, and renewable energy development have always been, in part, debates about the effect of these technologies on the environment. And the environmental effects of technology look quite different depending on the time scale being considered or the state of environmental knowledge when the analysis is done. For example, the environmental effects of refrigeration technology look much different now than they would have looked in an analysis done in the 1950s.

Research Needs

As with other human influences on the global environment, the effects of technology are likely to be contingent on the other driving forces. Consequently, research on the effects of technology on global change will need to consider the social context. Several critical topics for research are obvious: one involves comparisons of the environmental impacts of different technologies for energy production and consumption, food production, and other human activities that can have major impacts on the global environment, a topic that has received some attention in the past (e.g., Inhaber, 1978; Holdren et al., 1979, on energy production). Such studies should be specific at first, focusing on the alternatives available in a particular place and time, and should examine the technologies as they are implemented in actual social systems rather than under idealized conditions. Another involves diffusion of production technologies across national boundaries, particularly from more-developed to less-developed countries: How do the environmental impacts differ between the innovating countries and the adopting countries, and how do the differences depend on the social organizations using the technologies (e.g., Covello and Frey, 1989)? A third concerns the effect of government policies on the development, adoption, and use of technologies with different kinds of environmental effects: What policy choices influence technology and its use in environmentally destructive or beneficial ways, and how do the effects of policy depend on the political, economic, and social context where they are adopted (e.g., Zinberg, 1983; Clarke, 1988; Jasper, 1990)?

POLITICAL-ECONOMIC INSTITUTIONS

It seems reasonable that the social institutions that control the exchange of goods and services and that structure the decisions of large human groups should have a strong influence on the effects of human activity on the global environment. These institutions include economic and governmental institutions at all levels of aggregation.

A key institution is the market. Neoclassical economic theory argues that free markets efficiently allocate goods and services to the most valued ends. Thus, environmental problems can be analyzed in terms of market failures, that is, conditions that prevent markets from operating freely. Several types of market failure are relevant to environmental problems. First, the costs of the transactions necessary to resolve environmental problems in an optimal fashion may be prohibitively high because of the costs of collecting information, for example on the net present value to all affected of the future effects of resource use (e.g., Coase, 1960; Baumol and Oates, 1988). Second is the problem of "externalities." Individuals not involved in buying or selling a good or service may nevertheless be affected by the transaction, for example, if it alters the earth's ozone layer. But because they do not know what the effect will be, they may not engage in transactions to maximize their preferences. Third, government action may supersede the market (e.g., Burton, 1978; Coase, 1960), leading to inefficiencies, for instance, excessive and uneconomic cutting in U.S. national forests, or profligate use of coal in China due to artificially low prices and a production quota system that gives no premium for quality. Fourth, a lack of clearly defined private property rights may leave no one with the incentive to pay to prevent degradation. This situation can arise because of traditional social arrangements that allow free access to all (Hardin, 1968) or because of the indivisible, common-pool nature of resources such as open-access marine fisheries (Gordon, 1954) and the world atmosphere.

The analysis that traces environmental degradation to the absence of free markets is criticized on several grounds. First, even smoothly working markets are likely to produce undesirable outcomes. Questions have been raised regarding the theoretical assumption that a dollar has the same value regardless of a party's wealth and the morality of treating polluters and pollution recipients as symmetric and reciprocal sources of harm to one another (Kelman, 1987; Mishan, 1971). Second, the tendency of markets to place a higher value on possible impacts in the near future than on those in the distant future conflicts with the goal of long-term sustainability and reduces the rights of future generations effectively to zero (Weiss, 1988; Pearce and Turner, 1990). Third, goods that have no price, whose production is highly uncertain, or that are valued by nonparticipants in markets, for instance, the survival of nonhuman species, tend to be systematically undervalued in markets (e.g., Krutilla and Fisher, 1975). Fourth, the theory of market failures does not compare the environmental effects of different kinds of imperfect markets. Knowledge does not support the easy inference that the more a market resembles theoretical perfection, the more of the benefits of free markets it provides (Lipsey and Lancaster, 1956; Dasgupta and Heal, 1979). This is a serious limitation because, for environmental resources such as the stratospheric ozone layer, the only markets are imperfect.

Some analysts trace the roots of environmental problems to the system of free-enterprise competition that underlies markets (e.g., Schnaiberg, 1980). They argue that the capitalist, cash-based market system rewards those who exploit the environment for maximum short-term gain, an incentive structure fundamentally at odds with conservation and long-term sustainability and, more-over, that the capitalist class exacerbates the process through its strong influence on public policy. The argument is sometimes illustrated with the case of development in the Amazon.

The critique of capitalism can be criticized for relying on a global, highly generalizing contrast between capitalist market economies and precapitalist, subsistence, socially undifferentiated groups that presumably maintain a delicate balance with the natural environment. It does not account for the fact that noncapitalist societies without private property may perpetuate large-scale environmental abuses, as in the case of the drying of the Aral Sea for irrigation purposes in the Soviet Union (Medvedev, 1990) or the reliance on inefficient coal burning technology in China. It does not account for labor resistance to environmental protection when it seems to threaten loss of jobs, such as opposition to restrictions on mining and burning Appalachian coal. And it does not acknowledge the existence within fully integrated market economies of stable, intensively producing family farmers and smallholder land-use regimes that modify but do not permanently degrade their habitat.

Some analysts trace environmental deterioration, particularly in developing countries, to an international division between rich Western industrial and poor Third World raw material-producing nations that fosters political–economic dependence. Unequal terms of trade drain capital from peripheral or satellite regions to core areas. Underdevelopment and poverty are "developed" and perpetuated by market mechanisms (Wallerstein, 1976; Frank, 1967). This analysis emphasizes the effects of foreign investment, loans, the operations of large corporations, and quantifiable movements of capital, labor, imports, and exports on particular changes in the environment. Again, the Amazon case is sometimes offered as an example.

This dependency model highlights the important role of foreign capital and extractive industries, but because it pits a monolithic global capitalism against a similarly undifferentiated and largely passive Third World, it cannot account for the historical specificity of particular cases or the variability in internal dynamics as systems adapt (Wolf, 1982). Dependency theorists often overlook the role and complicity of national elites (Hecht and Cockburn, 1989). The model has been criticized as imprecise in that the notion of unequal terms of trade is inadequately defined. And contrary to the simple view of dependency, pressures from international lending institutions are now beginning to influence Amazonian land use in a positive way (Schmink and Wood, 1987:50; but see Price, 1989). Some Latin American countries, such as Costa Rica, have taken leadership in setting aside tracts of tropical forests as parks and conservation areas, despite high debt levels and dependence on exports to the United States (Gamez and Ugalde, 1988). A range of other factors in addition to dependency must be considered to account for the variety of resource use patterns in the Third World.

The state is a major institution affecting global environmental change because state actions modify economic institutions and affect a wide range of human actions, including those with global environmental impacts. As already noted, democratic states may be more responsive to popular pressures to take action on environmental problems than nondemocratic states. It may be more difficult in the latter for nonelite groups to get environmental issues on national policy agendas and then to influence the legislative process through the expression of public opinion. Another critical dimension may be the degree of centralization of the political system. One perspective argues that systems in which decisions are decentralized, primarily through markets, are apt to respond more readily to resource constraints. However, under certain circumstances, a more centralized, state-controlled form of decision making might be better able to take a long-term and broader perspective.

Specific public policies can also have significant environmental consequences, both intentionally and inadvertently. Many governments have pursued policies aimed at maximum exploitation of natural resources in pursuit of economic growth that give environmental concerns a low priority. However, many governments, primarily in the West, have also enacted policies to ameliorate the effects of industrial growth on the environment. State action can also have large unintended effects on the environment. For example, emissions of greenhouse gases and air pollutants in the United States have been greatly affected by the many policy choices of the U.S. government that have encouraged the use of the automobile as a form of personal transportation. Similarly, policies pursued by such federal agencies as the Army Corps of Engineers, the Department of the Interior, and the Atomic Energy Commission have affected environmental quality, even though—or perhaps because—environmental quality was not an issue in their policy deliberations. Knowledge about why different governments develop different environmental policies is discussed in more detail in Chapter 4.

Research Needs

Clearly, political–economic institutions can affect the global environment along many causal pathways. We have identified some of the important areas in which more knowledge could add greatly to understanding of the causes of global change. One is the comparative study of the effects of different imperfect-market methods of environmental management—including the various pricing systems and regulatory approaches in operation around the world, marketlike approaches not in use but potentially usable, and various mixtures—to determine their effects on global environmental variables as a function of where and when they are used, and on which human activities. Theoretical work classifying and analyzing the varieties of market imperfection could also make great contributions to understanding if directed toward the kinds of market imperfection characteristic of global environmental resources. A second important research area concerns the comparison of national policies in terms of their origins and their environmental effects. A third concerns the commonly alleged short-sightedness of corporate decisions about the environment. Under what conditions do capitalist actors adopt practices of natural resource use or waste management that preserve environmental values? What national policies affect the likelihood that they will adopt such practices? A fourth concerns the variation in development policies adopted by countries that are similar to each other in terms of level of development and dependency. To what extent is such variation dependent on the political structure of the state, national political culture, level of centralization of decision-making power, and other variables at the national level?

ATTITUDES AND BELIEFS

Widely shared cultural beliefs and attitudes can also function as root causes of global environmental change. Many analysts focus on broad systems of beliefs, attitudes, and values related to the valuation of material goods. An early argument in this vein attributed the modern environmental crisis to the separation of spirit and nature in the Judeo-Christian tradition (White, 1967); another traces the rise of capitalism with its materialist values and social and economic structures back to Protestant theology (Weber, 1958). The Frankfurt school of critical theory accorded a similar role to the spread of purely instrumental rationality (Habermas, 1970; Offe, 1985). Bias toward growth and a hubristic disregard for physical limits, others have argued, are today the principal driving forces (e.g., Boulding, 1971, 1974; Daly, 1977). Some point to "humanistic" values, derived from the Enlightenment, that put human wants ahead of nature and presume that human activity (especially technology) can solve all problems that may arise (Ehrenfeld, 1978). Some assert that increased environmental pressures are associated with materialistic values of modern society (e.g., Brown, 1981), implying that materialism is amplified in the social atmosphere of the Western world. Sack (1990) argues that environmental degradation is intimately tied to social forms and mechanisms that have divorced the consumer from awareness of the realities of production, hence leading to irresponsible behavior that exacerbates global change. And some analysts have traced environmental problems to a set of values, rooted in patriarchal social systems, that identify woman and nature and define civilization and progress in terms of the domination of man over both (e.g., Merchant, 1980; Shiva, 1989).

Some researchers argue that a secular change in basic values is occurring in many modern societies. Inglehart (1990) presents survey data to suggest that across advanced industrial societies, a value transition from materialist to postmaterialist values is occurring that has significant implications for the ability of societies to respond to global change with mitigation strategies that involve changes in life-style (see also Rohrschneider, 1990). Along a similar line, Dunlap and Catton have argued that a "dominant social paradigm" that sets human beings apart from nature encourages environmentally destructive behavior but that a "new environmental paradigm" that considers humanity as part of a delicate balance of nature is emerging (Dunlap and Van Liere, 1978, 1984; Catton, 1980; Catton and Dunlap, 1980). Other writers claim that a change in environmental ethics is necessary to prevent global environmental disaster (e.g., Stone, 1987; Sagoff, 1988).

Short-sighted and self-interested ways of thinking can also act as underlying causes of environmental degradation. The inexorable destruction of an exhaustible resource that is openly available to all, what Hardin (1968) called the "tragedy of the commons," is, at a psychological level, a logical outcome of this sort of thinking. Individuals seeking their short-term self-interest exploit or degrade open-access resources much faster than they would if they acted in the longer-term or collective interest (Dawes, 1980; Edney, 1980; Fox, 1985).

Direct challenges to these analyses are few, in part because they are compatible with analyses that emphasize the role of other driving forces. Cultural values, short-sightedness, and self-interestedness can both cause and respond to other major social forces, such as political–economic institutions and technological change. For example, global expansion of capitalism is seen by some as inextricably linked to a transformation of attitudes toward material production (Cronon, 1983; Merchant, 1991; Worster, 1988). Economists treat market behavior as an expression of preferences, which are ultimately attitudes, so the treatment of the environment is an indirect result of attitudes, even in economic analysis. Where controversy tends to arise is over the relative primacy and hierarchical ordering of attitudes and beliefs relative to other causal factors, especially the degree to which beliefs and attitudes can be given causal force in their own right or are products of more fundamental forces. The empirical associations underlying some claims have also been called into question (e.g., Tuan, 1968, on White, 1967). On the side of human response, however, at least some sense of the autonomy of attitudes and beliefs is implicit in every analysis that offers explicit recommendations for action.

Research Needs

As with the other driving forces, the most interesting questions for research concern the ways in which the central variables—here, cultural and psychological ones—interact with other driving forces to produce the proximate causes of global change. Observational and experimental studies of these relationships have been done, although almost always with relatively small numbers of individuals in culturally and temporally restricted settings (see, e.g., Stern and Oskamp, 1987, for a review). They indicate that attitudes and beliefs sometimes have significant influence on resource-using behavior at the individual level, even when social-structural and economic variables are held constant, and that attitudinal, economic, and other variables sometimes have interactive effects as well. But these studies do not explain the sources of variation in individual attitudes. It seems likely that attitudes and beliefs have significant *independent* effects on the global environment mainly over the long term—on the time scale of human generations or longer—and that within single lifetimes, attitudes function as intervening variables between aspects of an individual's past experience and that individual's resource use.

Testing this hypothesis would require research conducted over longer time scales than is common in psychological research.

CONCLUSIONS

This section distills some general conclusions or principles from the chapter and outlines their implications for setting research priorities.

THE PROXIMATE CAUSES

Research on the human causes of global environmental change should be directed at important proximate sources. It is critical to develop reasonably accurate assessments of the relative impact of different classes of human activity as proximate causes of global change. This chapter offers such an analysis—what we call a tree-structured account—for the human contribution to the earth's accumulation of greenhouse gases. Similar accounts should be made for the human contributions to other problems of global change. The task is relatively simple in the sense that the initial accounts need not have great precision. For social scientific work to begin, it will be sufficient to know whether a particular human activity contributes on the order of 20 percent, 2 percent, or 0.2 percent of humanity's total contribution to a global change. Such knowledge will allow social scientists to set worthwhile research priorities until more precision is available.

Current impact is not the only criterion of importance. Estimating the relative contributions of different future human activities to global changes is a more difficult, but equally important, part of assessing the importance of proximal causes. The difficulty lies in predicting future human activities, particularly the invention and adoption of new technologies. Initially, projections of the future accounts based on simple models will suffice to guide the research plan for human dimensions. However, researchers should be aware of their limitations and should occasionally test their analyses against a variety of scenarios of future human contributions to global change. Although it is more difficult to quantify other aspects of importance, these can provide strong justifications for research. For example, human actions that may be proximate causes of *irreversible* environmental change must be considered important beyond the magnitude of the change they may cause.

Researchers should be able to demonstrate the significance of their chosen subjects not only in terms of the theoretical and empirical issues in their fields, but also in terms of importance to global environmental change. All the research needs identified in this chapter presume that the importance criterion is applied to particular efforts to meet the needs.

Social Driving Forces

Understanding human causes of environmental change will require developing new interdisciplinary teams and will take lead time to build the necessary understanding. Listed below are some central considerations for guiding research.

The driving forces of global change need to be conceptualized more clearly. Different kinds of technological change and of economic growth clearly have different implications for the global environment, but much still needs to be learned about which aspects of change in these and other variables drive environmental change. A better typology of development paths is needed, so that researchers can identify the ways different styles of development affect the environment and the conditions under which a country or region takes one path or another. The same is true for research on the ways nation-states organize the management of natural resources.

Driving forces generally act in combination with each other. As the case studies demonstrate, the driving forces of global change are highly interactive. Brazilian deforestation is due to the combined effects of economic incentives, land tenure institutions, and government policy; Chinese coal use depends on the combined effects of economic development, the country's technological state, its political structure, and its economic policies; the development of CFCs was a function of population migration, economic incentives, and new technology. An additive model of these relationships is not viable, so the study of single causal factors in isolation is misleading.

The various driving forces should be studied in combination, using multivariate research approaches. These include quantitative multivariate studies that treat particular proximate causes (e.g., emissions of carbon dioxide and other greenhouse gases) as a joint function of population, economic activity, technological change, and political structures and policies. Such studies may be conducted using both national-level data on demographic and economic variables and indicators of policy and social-structural variables, some of which might have to be constructed for the purpose. Detailed case studies using qualitative methods are also important, as the case summaries in this chapter illustrate. Qualitative methods can offer a depth of understanding not available from quantitative analyses, which by their nature are limited to those variables already quantified. Moreover, each method acts as a check on conclusions drawn from the other.

Driving forces can cause each other. For example, new technologies can promote economic growth, which in turn allows for further technological development; materialistic ideologies contributed to the rise of capitalism, which promotes materialistic ideas. More complex mutual causal links also exist among several driving forces. Such relationships are difficult to disentangle and further complicate analyses of the human causes of global change. To understand the nature of these interactive relationships, it is important to compare different places and to follow the relationships over time.

The forces that cause environmental change can also be affected by it. Population growth is a good example of feedbacks between human actions and the global environment. Population growth increases the demand for food, which creates pressure to make agriculture both more intensive and extensive. These changes eventually bring diminishing returns, reducing food production per capita and creating downward pressure on population. The diminishing returns can be postponed by improved technology, but technology also interacts with the environment. Humans can increase food production by using tractor power, chemical fertilizers, pesticides, and herbicides, but these technologies rely on fossil energy and therefore eventually reach limits imposed by scarcity, price, or environmental consequences.

Relationships among the driving forces depend on place, time, and level of analysis. It is easy to illustrate the principle. For places: economic growth has been more dependent on fossil-fuel energy in China than in other countries, even other developing countries; the causes of deforestation in Brazil are distinct from its causes in other countries. For times: fossil-fuel energy use increased almost in lockstep with economic activity in industrialized countries for many years; since the 1970s, the correlation has been nearly zero (see Chapter 4). Also, the long-term effects on the global environment of a technology such as refrigeration with CFCs have been much different from the effects over a shorter time span—not only because of increasing use of the technology, but also because of the secondary effects of migrations made possible by the technology. For levels of analysis at the local level, the inefficiency of Chinese energy use can be understood in terms of outmoded technology and lack of funds for replacing it; at the national level, low prices for coal and the system of production quotas appear as critical factors; at the world level, the entire system of command economies is implicated. All the relationships are equally real and important, yet answers derived at each level are incomplete.

IMPLICATIONS FOR RESEARCH

1. *The highest priority for research is to build understanding of the processes connecting human activity and environmental change.* Better studies focused on the driving forces and their connections to the proximate causes are necessary for effective integrative modeling of the human causes of global change. Quantitative models will be of limited predictive value, especially for the decades-to-centuries time frame, without better knowledge of the processes.

More is generally known about the causes of population growth, economic development, technological change, government policies, and attitudes and culture—the driving forces of global change—than about their interrelationships and environmental effects. This is so because study of the driving forces is supported by organized subdisciplines or interdisciplinary fields in social science, such as population studies, development studies, and policy analysis, whereas *an interdisciplinary environmental social science—a field that examines the environmental effects of the driving forces—is not yet organized. There is a critical need for support of the research that would constitute that field.* Research on the processes by which human actions cause environmental change should begin from the basic principle that the relationships are contingent: the effect of such variables as population on environmental quality depends on other human variables that change over time and place. This fact has three major implications for research strategy: understanding the human causes is an intrinsically interdisciplinary project; the important human causes of global change are not all global; and comparative studies to specify the contingencies are critically needed (see #2 and #3 below). Research at the global level is important but far from sufficient for understanding the human causes of global change.

2. *Over the near term, research on the human causes of environmental change should emphasize comparative studies of glob-*

al scale. We can distinguish three types of global-scale analysis: aggregate, systemic, and comparative. Aggregate analysis at the global level examines human–environmental relationships on the basis of measures of the entire planet. Such analysis uses a small number of time-series data points and considers the entire planet the unit of analysis. For example, total atmospheric carbon dioxide can be correlated with global fossil fuel combustion over a period of time.

Systemic analysis of human–environmental relationships emphasizes facets of human activity that operate as a global system (i.e., a perturbation anywhere in the system has consequences throughout). For example, the world oil market is a global system in that changes in oil production anywhere reverberate through the system and may have global environmental impacts, for example, by changing the rate of consumption of oil or other fuels. Analyses of such relationships may use globally aggregate data or local and regional data linked to the phenomena of interest.

Comparative analysis at the global scale can take various forms. It might employ a large number of local or regional data points, worldwide in coverage. For example, the relationships of population, economic development, and government policies to deforestation may be studied by comparing data with the nation-state as the unit of analysis (e.g., Rudel, 1989). This approach is limited by the availability and comparability of relevant data (see Chapter 6). In contrast, case-based comparative studies can be selected so that a sample of units represents the range of socioeconomic and environmental contexts of the world. The case-comparison approach allows for more contextual detail at the expense of complete coverage. For example, a set of cases could be used to explore the various pathways that lead to conversion of wetlands to other uses.

Aggregate studies at the global level have limited value because the small number of data points make it impossible to identify the contingent relationships that shape the proximate human causes of global change. Systemic approaches have greater value in principle, but few human activities have the kind of systemic character that makes general circulation models of atmospheric processes valuable. Even the world oil market, one of the most globally systemic of environmentally relevant human systems, is affected by national policies such as trade restrictions and tax policies that interfere with world flows. Perhaps the most valuable research over the near term will come from comparative studies that involve either a large number of representative data points or a smaller number of selected regional case studies from around the world. The social sciences have a long tradition of comparative research and can usefully apply the conceptual and methodological tools they have developed to the problem of global environmental change.

3. *Human dimensions research should prominently include comparisons of human systems that vary in their environmental impact.* Comparisons between countries or localities or of the same place at different time periods can show why some social systems produce as much human welfare as others with less adverse impact on the global environment. A number of important issues lend themselves to comparative and longitudinal approaches, including:

—the causes of deforestation (studies can compare deforestation rates in countries that vary in their land tenure systems, development policies, and governmental structures);

—the effects of imperfect markets on release of greenhouse gases and air pollutants (studies can compare the emissions of countries or industries with different regulatory or pricing regimes);

—the sustainability of different agricultural management systems (studies might compare nearby localities in the same country);

—the effects of different industrial development paths on fossil fuel demand (studies might compare time-series data for different countries);

—the determinants of adoption of environmentally benign technologies or practices (for example, studies might compare industries or firms that do and do not recycle waste products);

—the relationship of attitudes about environmental quality and materialism to environmental policies in different countries.

Such studies can "unpack" broad concepts, such as technological change, economic growth, and population growth, that are frequently offered as explanations of how human activities cause global change. Comparative studies offer the best way to get inside the broad concepts and identify more specifically the features of growth and change in human activity that drive environmental change.

4. *Researchers should study the causes of major environmental changes both globally and at lower geographic levels.* Global aggregate analysis may show a very different picture from analyses at lower levels of aggregation. It is important to have both pictures because aggregate data can obscure the variety of causal processes that can produce the same outcome. For example, the global relationship between economic growth and greenhouse gas

emissions may change considerably if centrally planned economies become extinct. To estimate the size of any such effect, it is necessary to have studies at the national level. In addition, policy responses, particularly mitigation responses, require understanding of the activities that drive global change at the level at which the responses will be made. Depending on the topic, it may be important to conduct studies at the level of the nation-state, the community, the industry, the firm, or the individual. For studies below the global level, priorities should be set on the basis of the potential to gain understanding of the global picture or to make significant responses to global change. Thus, a high-priority study might be one that focuses on a country or activity that by itself contributes significantly to global change; or one that is expected to generalize to a sufficient number of individuals, firms, or communities to matter on a global level; or one that illuminates variables that explain important differences between actors at the chosen level of analysis. At each level of analysis, projects that meet such criteria are worthy of support, independently of what is known at the global level.

5. *Important questions should be studied at different time scales.* The full effects of technological and social innovations—both on society and on the natural environment—are often unrecognized for decades or centuries. The CFC case shows how the effects of human activities can look very different depending on the time scale used for analysis: a technology developed to refrigerate food had much wider global implications several decades later, after it was applied to refrigerating buildings. Such cases need to be collected so they can be studied systematically and testable hypotheses derived about what kinds of innovations are likely to acquire the social momentum that produces long-lasting and increasing effects on the global environment, such as has resulted from CFC technology or from the Brazilian development strategy used in the Amazon Basin. Theory is particularly weak for this purpose. Historians can offer convincing accounts of the current effects of changes of the distant past, but social scientists have little ability to project the effects of current changes in human systems equally into the future.

6. *Research should build understanding of the links between levels of analysis and between time scales.* For example, social movements mediate between individual attitudes and national policies; the interactions of individuals and firms can result in the creation of national and global markets; and national policies can stand or fall depending on whether thousands of firms or millions of individuals willingly comply. Because of these linkages, hu-

man action at one level of aggregation may depend on events at another level. Theory about these relationships is relatively weak, but the problem is of active interest to social scientists in several disciplines. If excellent data sets are compiled, the problem of connecting levels of analysis may attract leading disciplinary researchers to the topic of global change to build theory that would aid in understanding it while advancing their own fields.

Linking time scales is also critical to the global change agenda. The question is this: Which social changes, occurring on the time scale of months to years, are likely to persist or be amplified over time, to the extent that they will be significant to the global environment on a scale of decades to centuries? Obversely, which short-term changes are likely to disappear over time? Physical scientists know which halocarbons are long-lived catalysts for the destruction of stratospheric ozone and which ones are quickly destroyed; social scientists do not yet know much about which social changes catalyze other changes or about which ones are relatively irreversible. Historical cases, such as the CFC case, suggest some interesting hypotheses; over the near term, efforts to catalogue and compare such hypotheses would be a useful first step toward a theory of the long-term effects of social change. The general problem has received very little attention from social scientists. Improved understanding of the human analogues of long-lived catalysts may contribute to increased interest in long-term phenomena in social science.

NOTES

1 Some species, such as rosewood, are selectively eliminated from the forest for economic reasons. It is reasonable to expect that in an ecosystem characterized by many smaller species, such as insects dependent on a single species for food, that the selective cutting of one tree species will cause multiple extinctions.

2 The mechanism is rather complex. Evapotranspiration in the Amazon forest appears to cause a regional climatic increase in precipitation. In such a regime, large-scale clearing, which reduces evapotranspiration per land area even if trees are replaced by other vegetation, will decrease rainfall downwind. Because species diversity in Amazonia is directly related to levels of rainfall, lower rainfall in any region can be expected to reduce the number of species in that region.

3 Species with large area requirements are disproportionately affected when forest clearing is fragmented, as it typically is in Amazonia. Under those conditions, an individual or functional group of individuals with a large area requirement is less likely to find adequate forest resources

within its area. Species with wide ranges are unlikely to be extinguished by habitat destruction within their range, but such destruction is likely to eliminate entirely the habitats of some of the species in the area with smaller ranges. Finally, although humans might be expected to husband populations of species with economic value, this has not typically been the case on frontiers, as the exploitation of Amazonian rosewood and the American bison illustrate.

4
Human Consequences and Responses

Since before recorded history, environmental changes have affected things people value. In consequence, people have migrated or changed their ways of living as polar ice advanced and retreated, endured crop failures or altered their crops when temperature and rainfall patterns changed, and made numerous other adjustments in individual and collective behavior. Until very recently, people have responded to global phenomena as if they were local, have not organized their responses as government policies, and have not been able to respond by deliberately altering the course of the global changes themselves. Things are different now from what they have been for millennia.

This chapter examines the range of human consequences of, and responses to, global environmental change. We begin by developing the concept of human consequences and showing why, to understand them, it is critical to understand the variety of human responses to global change. We then offer a framework for thinking about human responses and discuss the pivotal role of conflict. The next section examines three cases that illustrate many of the major factors influencing the human consequences of global change. The following sections describe the human systems that are affected by or respond to global change, and how they interrelate. We conclude by offering some general principles for research and some research implications.

UNDERSTANDING HUMAN CONSEQUENCES

Many human actions affect what people value. One way in which the actions that cause global change are different from most of these is that the effects take decades to centuries to be realized. This fact causes many concerned people to consider taking action now to protect the values of those who might be affected by global environmental change in years to come. But because of uncertainty about how global environmental systems work, and because the people affected will probably live in circumstances very much different from those of today and may have different values, it is hard to know how present-day actions will affect them. To project or forecast the human consequences of global change at some point in the relatively distant future, one would need to know at least the following:

—the future state of the natural environment,
—the future of social and economic organization,
—the values held by the members of future social groups,
—the proximate effects of global change on those values, and
—the responses that humans will have made in anticipation of global change or in response to ongoing global change.

These elements form a dynamic, interactive system (Kates, 1971, 1985b; Riebsame et al., 1986). Over decades or centuries, human societies adapt to their environments as well as influence them; human values tend to promote behavior consistent with adaptation; and values and social organization affect the way humans respond to global change, which may be by changing social organizations, values, or the environment itself.

This complex causal structure makes projecting the human consequences of global change a trickier task than is sometimes imagined. It is misleading to picture human impacts as if global change were like a meteorite striking an inert planet, because social systems are always changing and are capable of anticipation. So, for example, an estimate of the number of homes that would be inundated by a one-meter rise in sea level and the associated loss of life and property may be useful for alerting decision makers to potentially important issues, but it should not be taken as a prediction, because humans always react. Before the sea level rises, people may migrate, build dikes, or buy insurance, and the society and economy may have changed so that people's immediate responses—and therefore the costs of global change—may be different from what they would be in the present.

One may imagine human consequences as the output of a matrix of scenarios. Assume that four sets of scenarios are developed for the futures of the natural environment, social and economic organization, values, and policies. Joining together all combinations of one scenario from each set, and adding assumptions about people's immediate responses, would generate an extensive set of grand scenarios. The human consequences of global change could then be defined as the difference between the state of humanity at the end of one grand scenario and the state of humanity at the end of a base case or reference scenario with a different natural-environment component. By this definition, a particular change in the natural environment has different consequences depending on the scenarios assumed for society, values, and responses.

Building these scenarios, identifying the most probable ones, and assessing their outcomes would be an overwhelming analytic task. Rather than trying to set a research agenda for that task, we undertake in this chapter a less demanding but still very difficult task: to focus on human responses to global change broadly conceived. We do not discuss ways to improve forecasts of the state of the natural environment; that topic is outside the range of human dimensions. Neither do we devote much attention to improving forecasts of social and economic organization or of human values, even though these topics clearly belong to the social sciences and are critical to understanding the effects of global change. We bypass these issues because the need for improved social, economic, and political forecasting is generic in the social sciences, and addressing this broad need would take us far beyond our charge to focus on human–environment interactions. We offer only limited discussion of how future global change might proximally affect what humans value, because the variety of possible global changes and the uncertainty about the effects of each make it far too difficult to go into detail. Instead, we review basic knowledge about how human systems respond to external stresses, in the context of discussing human responses.

In our judgment, understanding human responses is key to understanding the human consequences of global change. We do not mean to downplay the importance of certain kinds of research that do not focus explicitly on responses. Two such research traditions, in particular, are highly relevant. The impact-assessment tradition involves projecting the human consequences of a range of natural-environment scenarios under given assumptions about human response. The tradition of post hoc case analysis involves assessing the actual human outcomes after past environmental changes (and given the responses that actually occurred), in the hope of drawing more general conclusions. Research in these traditions, combined with analysis of human response, can offer valuable insights into the human consequences of global change. We discuss that research as appropriate in this chapter and in Chapter 5.

SOME DIMENSIONS OF HUMAN RESPONSE

The human responses relevant to global change differ along several dimensions. We consider the following analytic distinctions useful for thinking about the range of responses available.

Responses to Experienced Versus Anticipated Change

People and social institutions may respond to environmental change as it is experienced (post facto) or as it is anticipated.[1] In the past, people responded mainly to experienced environmental change; only in very recent history, because of increasing scientific knowledge, has there been any rational basis for anticipatory responses. Policy makers and others are now faced with a variety of options, some of which involve anticipatory action and some of which depend on awaiting the experience of global change.

Deliberate Responses Versus Actions with Incidental Effects

Some human actions can be taken deliberately in response to global change. For instance, people can build dikes to keep out rising seas or reduce greenhouse gas emissions to mitigate global warming. Human actions can also affect human responses to global change incidentally to their intended purposes. For example, European settlement of the Americas gave Europeans and, later, others access to a wider variety of food crops, making human survival less dependent, at least in principle, on a small number of staples that might be vulnerable to altered growing conditions caused by environmental change. World markets have subsequently reduced the number of major staple foods so that, in practice, people may eat no larger a variety of foods than before (Plotkin, 1988). High taxes on gasoline in Europe and Japan, enacted for reasons unrelated to the global environment, encouraged development and purchase of small, fuel-efficient automobiles that incidentally slow the pace of global warming. By bringing about technological change, these taxes also incidentally have helped make it easier for all countries—even those without high gasoline taxes or companies that produce fuel-efficient automobiles—to respond to the challenge of global warming with improved energy efficiency.

Changes in society that incidentally affect human responses to global change are important both directly and because they could become tomorrow's deliberate responses. For example, gasoline taxes, which were not initiated with the global environment as a consideration, could be increased to cut CO_2 emissions. Studies of the incidental effects of such actions might inform decision makers about what could happen without deliberate intervention and about which present policies might make societies more robust in the face of global change. Both kinds of knowledge are essential for informed policy debates.

Coordinated Versus Uncoordinated Responses

Response to global change may be coordinated, as through the policies of governments or trade associations aimed at eliciting the same action from many actors, or uncoordinated, as with independent actions of households or small firms. Both types of response can be either anticipatory or post facto; both can affect global change either deliberately or incidentally. Moreover, coordinated and uncoordinated responses can be connected to each other, in that coordinated actions by governments and industries can create new options for uncoordinated actors, prohibit responses, or raise or lower their costs.

Interventions at Different Points in the Process

Figure 4-1 elaborates on Figure 2-2 to show how human action can intervene at any point in the cycle of interaction between human and environmental systems to protect against threats to what humans value. We offer the following rough distinctions among types of interventions.[2]

The term *mitigation* is generally used to describe interventions on the human causes side of the diagram. Mitigation includes all actions that prevent, limit, delay, or slow the rate of undesired impacts by acting directly or indirectly on environmental systems. Mitigation can operate at various points in the causal cycle.

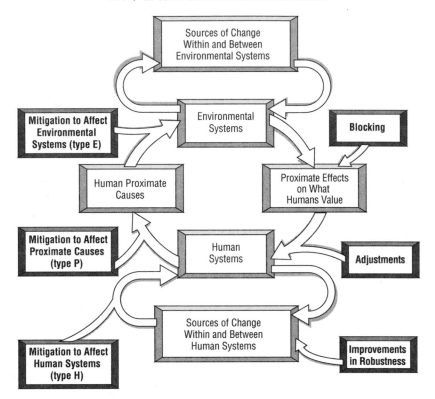

FIGURE 4-1 Interactions between human and environmental systems and the role of various types of human response. Lightly shaded boxes repeat the relationships presented in Figure 2-2.

It may involve *direct interventions in the environment* (type E in the figure) to counteract the effects of other human actions, *direct interventions in the proximate human causes* (type P), and *interventions in the human systems* (type H) that drive global change, intended to have an indirect or downstream effect on the proximate causes.

For example, global warming is the direct result of a change in the earth's radiative balance; humans can mitigate global warming by any actions that slow the rate of change or limit the ultimate amount of change in the radiative balance.[3] They can intervene in the environment (type E), for example by directly blocking incident solar radiation with orbiting particles or enhancing the ocean sink for carbon dioxide by adding nutrients. They can intervene in the proximate causes (type P), by regulating automobile use or engine design to cut carbon dioxide emissions or limiting the use of certain nitrogen fertilizers to reduce nitrous oxide emissions. They can intervene in human systems (type H) and indirectly control the proximate causes, by investing in research on renewable energy technologies to replace fossil fuel or providing tax incentives for more compact settlements to lower demand for transportation.

Mitigation of ozone depletion might, in principle, involve release of substances that interact chemically with CFCs, producing compounds with benign effects on the stratospheric ozone layer (type E), limiting emissions of chlorofluorocarbons (CFCs) and other gases that deplete ozone (type P), or developing alternative methods of cooling buildings that do not rely on CFCs (type H). Mitigation of threats to biological diversity might include, at least in principle, engineering new varieties, species, or even ecosystems to save diversity, if not individuals (type E); limiting widespread destruction of tropical forests, estuaries, and other major ecosystems (type P); or promoting systems of land tenure and agricultural production that decrease the pressure for extensive development of tropical forests (type H).

Humans can intervene in several ways on the response side of the cycle. Such actions are sometimes generically called *adaptation*, but there are important distinctions among them. One type of response, which can be called *blocking*, prevents undesired proximate effects of environmental systems on what humans value. It can be described by example. If global climate change produces sufficient warming and drying (drought) on a regional scale, it may threaten the region's crops; development and adoption of drought-resistant crops or crop strains can break the connection between environmental change (drought) and famine by preventing crop failure. Similarly, loss of stratospheric ozone threatens light-skinned humans with skin cancer, through exposure to ultraviolet radiation; avoidance of extreme exposure to sun and application of sunscreens help prevent cancer, although they do not mitigate the destruction of the ozone layer. Tropical deforestation threatens species with extinction by eliminating their habitats; creation of forest preserves would provide many species sufficient habitat to survive, while doing little to slow net deforestation.

Another type of adaptive response is to prevent or compensate for losses of welfare that would otherwise result from global change. Such actions can be called *adjustments*.[4] They neither mitigate environmental change nor keep it from affecting what people value, but rather intervene when a loss of welfare is imminent or after it has begun to be manifest. Examples include evacuation from areas stricken with flood or drought, food shipments or financial assistance to those remaining in such areas, and development of synthetic substitutes for products previously obtained from extinct species.[5]

Yet another type of response, sometimes called anticipatory adaptation, aims to *improve the robustness* of social systems, so that an unchecked environmental change would produce less reduction of values than would otherwise be the case.[6] This type of intervention does not alter the rate of environmental change, but it lowers the cost of any adjustments that might become necessary. It can be distinguished, at least in theory, from type H mitigation in that it does not necessarily alter the driving forces of global change. An example is diversification in agricultural systems. Farmers, regions, and countries that rely on a range of crops with different requirements for growth may or may not produce less greenhouse or ozone-depleting gases than monoculturists. But polycultures are more robust in the face of drought, acid deposition, and ozone depletion. There may be crop failure, but only in some crops. Similarly, families and communities that have both agricultural and nonagricultural income are harmed less by the same threats than purely agricultural groups. They have other sources of income and can purchase crops from elsewhere.[7]

All social systems are vulnerable to environmental change, and modern industrial societies have different vulnerabilities from earlier social forms. Modern societies have built intricate and highly integrated support systems that produce unprecedented material benefits by relying critically on highly specialized outputs of technology, such as petrochemical fertilizers and biocides; hybrid seeds; drugs and vaccines; and the transmission of electricity, oil, and natural gas from distant sources. Although these complex sociotechnical systems contain great flexibility through the operation of global markets, they may have vulnerabilities that reveal themselves in the face of the changes that these systems have helped create. For instance, modern societies have become highly dependent on fossil fuels and vulnerable to a serious disruption of supply or distribution systems. They also support much larger and denser populations than ever before; such populations may be vulnerable to ecological changes affecting the viability of their food supplies.

Evidence from studies of disasters suggests that the poor, who lack diversified sources of income, political influence, and access to centralized relief efforts, tend to be worst off (Erikson, 1978;

Kroll-Smith et al., 1991; Mileti and Nigg, 1991). However, studies to assess the vulnerabilities of larger human systems, such as national or world food or energy systems, are rarely done (e.g., Rabb, 1983). The far side of vulnerability is also little studied: When a system fails to resist environmental pressure, under what conditions does it return to its previous state? If it undergoes permanent change, what determines the nature of the new state?

THE PIVOTAL ROLE OF CONFLICT

An important consequence of global environmental change is conflict, because global change affects what humans value, and different people value different things. When U.S. energy use threatens the global climate or land clearing in Brazil threatens the extinction of large numbers of species, people around the world are understandably concerned. They may express a desire—or even claim a right—to influence the choices of people or governments continents away. And the people or countries subjected to those claims may resist, especially when they feel that changing their behavior will mean suffering. The further global change proceeds, the more likely it seems that it will be a source of conflict, including international conflict, over who has a right to influence the activities implicated as causes, who will pay the costs of responding, and how disputes will be settled.

A Current Controversy: To Mitigate or Not to Mitigate?

One of the most heated policy debates about responses to a global change is between advocates of immediate efforts to mitigate global warming and those who would postpone such action. This debate arose within the committee, even though we were not charged with recommending strategies for response to global change. We offer the following brief, sharply stated version of the debate to highlight some important characteristics of controversies about global change: that they are partly, but not entirely, fact-based; that they are likely to persist even in the face of greatly increased knowledge about the causes of global change; and that they are pervasive, even in discussions restricted to research priorities.

In one view, the wise course of action on global warming is to conduct research on the phenomenon but not to take action to slow or mitigate it until the phenomenon is better understood. Proponents of this view make the following arguments:

1. *Uncertainty of global change.* The nature and extent of global warming in the future is highly uncertain because of incomplete knowledge of the relevant properties of the atmosphere, oceans, biosphere, and other relevant systems. It is wasteful for society to expend resources to prevent changes that will not occur anyway. Moreover, the mitigation efforts may themselves set in motion undesired changes.

2. *Adjustment will make mitigation unnecessary.* Human systems can adjust to global climate changes much faster than they are likely to occur. The projected doubling of atmospheric carbon dioxide levels will take place about 80 years from now. By contrast, financial markets adjust in minutes, administered-market prices in weeks, labor markets in years, and the economic long run is usually reckoned at no more than two decades. The implication for action is that what individuals and organizations do on their own in anticipating climate change may be sufficiently successful that organized, governmental responses will be superfluous. The impact of climate change will reach people through slow price increases for the factors of production; in reasonably well-functioning markets, economic actors adapt readily to such changes. They invent industrial processes that economize on scarce inputs, find substitutes, purchase energy-efficient equipment when energy prices are rising, and so forth. In the past, such adjustments have contributed to human progress, and there is every reason to expect that pattern to continue.

3. *Don't fight the wrong war.* It makes no sense to act like the generals who built the Maginot Line for the wrong war or to construct dikes for cities whose populations will have moved or dams to water crops that will be grown elsewhere. Technological and social changes often eliminate problems without any specific mitigation efforts by changing the offending technology or making it obsolete. For example, boilers no longer explode on trains because they no longer use steam engines; horses are no longer the main polluters of urban streets. Concern about the greenhouse effects of fossil fuel burning will prove premature if development of fusion or solar energy technology can replace most fossil fuel use over the next 50 years.

4. *Better policy options may lie on the horizon.* Further research may identify more effective and less costly interventions than those now available. For example, it has recently been suggested that adding iron to the oceans to fertilize phytoplankton that would absorb carbon dioxide from the atmosphere may be a way to address the greenhouse effect (Martin et al., 1990). That

proposal, whatever its ultimate feasibility or desirability (Lloyd, 1991), demonstrates that improved understanding of biogeochemical systems might generate promising new proposals for mitigating global change. Improved understanding of social systems has reasonable potential to discover other classes of effective response.

5. *It may be more costly to act now.* Actions that can be postponed will be less burdensome because of continuing economic progress. If people living in the 1890s had invested in preventing today's environmental problems, their expense on our behalf would probably have been made on the wrong problems, and it would have been an inequitable transfer of resources from a poorer generation to a richer one. It probably makes no more sense for the current generation to sacrifice to benefit a future, even wealthier generation. This is the argument for a positive social discount rate. It assumes that expenditures made now could otherwise be invested at compound interest in improvements in human well-being. If the growth rate for such investment exceeds the average rate at which environmental problems develop, people will be better off in the future if they do not spend on mitigation now.

Proponents of immediate mitigative action make the following arguments:

1. *Action now is more feasible and effective than action later.* It is in the nature of exponential growth processes that the earlier the growth rate decreases, the greater the final effect. Bringing down the birth rate in India to two children per couple in 1995 rather than in 2005 can make a difference of 300 million people by the time the Indian population stabilizes (Meadows, 1985). To achieve the same effect by starting later would impose greater restrictions on the people living at that time. It is therefore easier to mitigate the effects of exponential growth the sooner the effort is made.

2. *It is easier to adjust to slower change.* Mitigation is prudent because of the long time lags in the global environmental system. By the time it becomes clear that a response is needed, it may be too late to prevent catastrophe if the change is proceeding rapidly. Even if catastrophe is unlikely, mitigation that slows the rate of change makes it more likely that adjustments can be made in time. This is clearly the case for nonhuman organisms, such as tree species that can adjust to climatic change by migrating, as seedlings move to more favorable locations. Such species have a

maximum rate of migration, so can adjust to climatic change below that rate.[8] The same principle probably also applies to human adjustments to major environmental change.

3. *It is wise to insure against disaster.* Mitigation in the face of possibly catastrophic outcomes is like taking out insurance against flood and fire. The insurance expenses are bearable, but the expenses of catastrophe may not be.

4. *Avoid irretrievable error.* It is wise to mitigate against potentially irretrievable losses. The clearest example is species extinction. If species are valued for themselves, their loss is irretrievable; even if they are valued only for what benefits they may have for humanity, species loss may be irretrievable. Other environmental values, such as loss of the life-supporting capacity of wetlands or large bodies of water, may also be irretrievable; often we do not know until the values are lost.

5. *Avoid high-risk environmental experiments.* Humans are now conducting large-scale uncontrolled experiments on the global environment by changing the face of the earth and the flows of critical materials at unprecedented rates. It is prudent to limit the pace and extent of such experiments because of the likelihood of unanticipated consequences. Like natural mutations, most of these experiments are probably destined to fail, and there is only one global environment to experiment on. As the extent of human intervention in the global environment continues to increase, so does the strength of this argument. The argument supports mitigation efforts that slow ongoing human interventions in the environment, but generally not those that would stop greenhouse warming by new interventions in the global environment.

6. *Economic arguments do not encompass some environmental goods.* The discount-rate argument is specious in the general case because the costs and benefits of postponing action are not always commensurable. Some important and meaningful trade-offs can be made on economic grounds, for instance, between investing in renewable energy development and in directly limiting the burning of fossil fuels. But sometimes the economic logic makes no sense. If current economic activity destroys the life-support systems on which human life depends, what investment at compound interest could ever recoup this cost? Economic arguments also cannot deal with some things—including the balance of nature—on which people place intrinsic or spiritual value. To the extent people want to preserve such values, mitigation is the only acceptable approach. Moreover, economic accountings systematically undervalue things—such as genetic resources—for

which there are few property rights or for which economic value is only potential.

7. *Some mitigative action is fully justified on other grounds.* A good example is investments in energy efficiency that provide an excellent return on investment even with narrow economic calculations. Such actions can achieve the benefits of mitigation at no extra cost, while providing other benefits.

Implications of Conflict About Human Response

Many controversies are beginning to develop out of concerns with global change. One pits Third World countries against the developed countries that are now becoming concerned with limiting use of fossil fuels and restricting the felling of tropical forests. The Third World position, of course, is that other countries used fossil fuels and undeveloped frontiers for their economic development, and fairness dictates that the poorer nations now have their turn. Many analysts believe that if large-scale climate change results from human activities, the poorer countries are likely to suffer most because they lack resources they could use to adapt. Such an outcome would produce yet other conflicts.

The controversies about global change are only partly fact-based. True, some of the disagreements might fade with better knowledge about the global environment and the likely effects of different feasible responses. As it became clear that expected global warming over the next 50 years could not cause the breakup of the West Antarctic icecap, the flood-prevention rationale for slowing greenhouse gas emissions became considerably weaker. A response such as dike building seems much more appropriate when the sea threatens only a few areas. And if it became clear what each policy option—at the local, national, and international levels—would accomplish if enacted, some of them could easily be rejected.

But knowledge often fails to resolve controversy. It frequently raises new disputes or calls old beliefs into question. And even when new knowledge reduces uncertainty, controversies persist because not only facts, but also important interests and values, are at stake. Informed people disagree because the remaining uncertainty leaves room for judgment, because they may assume different scenarios about the future of society, and because an outcome that harms what one person values may enhance what another values. Those impressed with the potential benefits of economic growth tend to line up against those who fear of the potential costs; those with a strong faith in the ability of human ingenuity to solve life's problems line up against those awed by what is at stake; those who stand to benefit from an outcome line up against those who stand to lose. When faced with choices, some prefer international solutions to global problems, others see national action as more feasible; some favor market adaptations, others, community-based action outside the market and the state; some are attracted to large-scale technological solutions, others see them as cures that may be worse than the disease. In short, the debates are not only about the workings of human and environmental systems, but also about political and economic interests, conflicting values and faiths, differing assumptions about the future, and different judgments about resiliency in the face of the unexpected.

Research on Conflict Studies of environmental and technological conflict are a significant part of social research on conflict (e.g., Nelkin, 1979; Mazur, 1981; Freudenburg and Rosa, 1984; Jasper, 1988; Clarke, 1989). Issues of global environmental change have all the features characteristic of the most difficult technological controversies: awareness of human influence on the hazards, serious worst-case possibilities, the possibility of widespread and unintended side effects, delayed effects not easily attributable to specific causes, and lack of individual control over exposure (National Research Council, 1989b:57-62).

Social science can help illuminate the nature of environmental controversies and evaluate ways of managing them. Social scientists interested in environmental policy have studied the conditions shaping and favoring the resolution of environmental controversies and the role of scientific, governmental, and mass media communication in the decision process (e.g., Dietz and Rycroft, 1987; Gould et al., 1988; Jasanoff, 1990; Nelkin, 1979, 1988; National Research Council, 1989b). Some have begun to consider the various ways environmental change might lead to conflicts with the potential for violence (e.g., Homer-Dixon, 1990).

Social scientists specializing in conflict have developed generalizations that might be more thoroughly applied to environmental conflict. For example, conflicts may be based mainly on ideology, interest, or understanding (Aubert, 1963; Glenn et al., 1970; Rapoport, 1960, 1964; Hammond, 1965; von Winterfeldt and Edwards, 1984; Syme and Eaton, 1989), and different types of conflict tend to yield to different tactics of resolution (e.g., Druckman and Zechmeister, 1973; Druckman et al., 1977). Defining an environmental conflict as either one of understanding or one of interests and values affects which groups and arguments are considered legitimate in policy debates (Dietz et al., 1989). The nature of the relationship between the parties to a conflict can determine whether the conflict focuses on ideological positions (e.g., Campbell, 1976; Zartman and Berman, 1982), differences in understanding (e.g., Axline, 1978), or differences in interests (e.g., Strauss, 1978). And the behavior of the parties to a conflict depends on the pattern and relative strength of incentives to compete and to cooperate (e.g., Pruitt and Kimmel, 1977), the probability of continued interaction in the future (e.g., Axelrod, 1984), and on whether two or more parties are involved (Groennings et al., 1970; Hopmann, 1978; Putnam, 1988).

More research seems warranted to use existing knowledge about conflict to illuminate the ways social conflict may result from global environmental change. This research would investigate the ways environmental changes may affect organized social groups and their resource bases and would hypothesize links between those effects and conflict. A first step is to construct an analytical framework for identifying the possible routes from particular environmental changes to particular types of conflict. The framework of Homer-Dixon (1990) provides a start, for causes of violent conflict. Case analyses of past social conflicts can be used to assess hypotheses drawn from such analytic frameworks.

Research on Conflict Resolution and Management Social scientists have also identified a number of approaches for resolving or managing policy disputes, some of which are beginning to be studied in the context of environmental conflicts. These include mediation techniques intended to address the value dimension of environmental conflict (e.g., Ozawa and Susskind, 1985); facilitation procedures that emphasize problem-solving discussions and have proved useful as a prelude to negotiation (Burton, 1986; Druckman et al., 1988); techniques of separating values from interests to makes conflicts appear smaller and easier to solve (Fisher, 1964; but see Druckman, 1990); efforts to focus on shared principles for decisions (Zartman and Berman, 1982) or to discuss values as ranked priorities rather than ideological differences (Seligman, 1989); policy exercises that emphasize creative use of scientific knowledge to solve environmental problems (Brewer, 1986; Toth, 1988a, b); and computer software for dealing with the cognitive and political aspects of both conflicts over the interpretation of data for environmental management (Hammond et al., 1975; Holling, 1978).

The nature of technological conflicts suggests, however, that over the long term, management is a more realistic goal than stable resolution. Recent work on risk communication is potentially relevant to social responses to global change because global change problems, like those to which that literature refers, are characterized by high levels of scientific uncertainty and great potential for conflict about social choices (Covello et al., 1987; Davies et al., 1987; Fischhoff, 1989; National Research Council, 1989b; Stern, 1991). This work suggests that institutions responsible for decisions about global change will also have to manage conflict. These institutions will need to provide accurate information, but should not expect information to resolve conflict. The institutions will need to make a place for the stakeholders to be represented from the earliest stages of the decision process, ensure openness in processes of policy decision, include mechanisms for the main actors to have access to relevant information from sources they trust, and use the conflicting perspectives and interpretations of current knowledge and uncertainty to inform the ongoing debate (National Research Council, 1989b; Stern, 1991).

Research Needs Relatively little is known about the structure of particular conflicts about global change at the local, national, and international levels or about which means will be most effective in dealing with them. Therefore, we recommend increased empirical research, including both field studies and laboratory-simulation studies, to clarify the sources and structures of particular environmental conflicts and to test the efficacy of alternative techniques for their resolution and institutions for their management.

HUMAN RESPONSE: THREE CASES

In Chapter 3 we presented cases to illustrate how human actions can contribute to the causes of global change. Here we present three cases to illustrate the human consequences of, and responses to, environmental change. Taken together, they show the importance of all the major human systems involved (described later in the chapter) and the ways that conflicts are played out and choices made within these systems.

International Regulation of Ozone-Depleting Gases

As mentioned earlier, the most successful effort to date to address a global environmental problem by international agreement is the ozone regime, articulated in the 1985 Vienna Convention for the Protection of the Ozone Layer, the 1987 Montreal Protocol on Substances That Deplete the Ozone Layer, and the 1990 London Amendments to the protocol. This regime, in its current form, commits its members to phasing out the production and consumption of CFCs and a number of related chemicals by the year 2000. The regime represents the first concerted international effort to mitigate "a global atmosphere problem before serious environmental impacts have been conclusively detected" (Morrisette, 1989:794).

The political history of the ozone regime begins as a national issue in the United States and a handful of other Western countries in the early 1970s, in connection with emissions from supersonic transport (SST) aircraft and then from aerosol spray cans (Downing and Kates, 1982; Morrisette, 1989). Environmental groups organized opposition to the development of the SST and to the extensive use of aerosols. Individual responses led to a sharp drop in sales of aerosol products (Morrisette et al., 1990). The U.S. Congress, prodded by government studies supporting the CFC-ozone depletion theory and its links to skin cancer, approved the Toxic Substances Control Act of 1976, which among its other provisions, gave the Environmental Protection Agency (EPA) the authority to regulate CFCs. In 1978, the United States became the first country to ban the nonessential use of CFCs in aerosols. However, the EPA ruled that other uses of CFCs, such as in refrigeration, were both essential and lacked available substitutes.

Ozone depletion emerged as a major international issue in the 1980s. This occurred primarily as a result of initiatives by the United Nations Environment Programme (Morrisette, 1989) and the actions of the international scientific community (Haas, 1989), with the support of the international environmental movement. The Vienna convention of 1985 embodied an international consensus that ozone depletion was a serious environmental problem. However, there was no consensus on the specific steps that each nation should take.

A number of events in 1986 and 1987 created a new sense of urgency about the depletion of stratospheric ozone. These included a rapid growth in demand for CFCs due to new industrial applications and the end of a global economic recession; important new studies by the World Meteorological Organization, the National Aeronautics and Space Administration, the Environmental Protection Agency (EPA), and the United Nations Environment Programme; and, most important, the widely publicized discovery by scientists of the Antarctic ozone hole in 1985. In January 1986, EPA initiated a series of workshops designed to build an international scientific consensus supporting the need to control the use of CFCs. In the same year, DuPont announced that its scientists had determined that CFCs were the most likely cause of ozone depletion. These events persuaded American officials of the need for decisive international action. When negotiations on a protocol to the Vienna convention for controlling CFCs resumed in December 1986, the United States adopted a firm position, calling for an international treaty not only freezing production of CFCs but also reducing production and consumption.

Following extensive and complex negotiations, the Europeans, whose earlier opposition to a cutback in production had prevented agreement in Vienna, moved closer to the U.S. position. They were persuaded to do so by three factors: the weight of scientific evidence, pressures from their own domestic environmental groups, and the fear that, in the absence of a treaty, the United States might take unilateral action to impose trade sanctions. While compromises on several controversial points proved sufficient to gain Japanese and Soviet adherence, the major developing countries (e.g., China and India) did not become signatories to the Montreal Protocol.

Only after the Montreal Protocol was signed did the full extent of ozone depletion became public: ozone depletion over Antarctica reached a historic high in 1987, and the link to the release of CFCs became a matter of scientific consensus. DuPont responded by announcing that it planned to discontinue CFC production by the end of the century and, in March 1989, 123 countries called for the absolute elimination of production by the same date. A resolution agreeing to totally phase out all production and consumption of CFCs by the year 2000 was adopted by 81 countries in May 1989 at the first governmental review of the Montreal Protocol.

Taking advantage of this momentum, the parties to the Montreal Protocol, meeting at a review conference in London in June 1990, were able to negotiate a series of strong amendments. These amendments accelerate the phaseout schedule for CFCs and halons and add methyl chloroform and carbon tetrachloride to the list of chemicals to be eliminated. Equally important, the amendments establish an international fund to be used to assist developing countries in switching to substitutes for CFCs in the production of refrigerators and air conditioners. On the strength of this development, both China and India agreed to become members of the international ozone regime.

Why was it possible to reach a broad international agreement restricting CFCs? Analysts have identified four important factors: an evolving scientific consensus; a high degree of public anxiety in developed countries about the risks associated with the continued use of CFCs, due in large measure to an association with skin cancer; the exercise of political muscle by the United States; and the availability of commercial substitutes for CFCs (Haas, 1989; Morrisette, 1989). The last served the critical role of diminishing the opposition of the chemical industry to a phased reduction. When DuPont, the producer of 25 percent of all CFCs, decided to develop substitutes, it "forced other CFC manufacturers to follow suit or risk losing market share" (Haas, 1989:11). Haas adds that, because this issue could be resolved by a technical fix, it did not involve any hard choices and therefore may be unique in the annals of global environmental change.

Another important influence in getting CFCs on political agendas may have been the efforts of the scientific community, which has been influential in drawing attention to other environmental problems (Haas, 1991). Haas (1989) notes that it was initially a group of atmospheric physicists and chemists, most of whom worked in the United States, who attempted to place the issue of ozone depletion on the national and global environmental agendas, and that this community continued to press the issue throughout the 1980s. He argues that the speed of policy response in the United States may have been due to the "highly fragmented nature of American government and society [which] facilitates access of a strongly motivated group of technical experts" (p. 8). Thus, the access of a key group to policy debates at the national level may have influenced international action on CFCs.

The history of the ozone regime illustrates a number of key variables that affect the likelihood of reaching similar agreements on other global environmental problems (Sand, 1990b; Benedick, 1991). Further studies are desirable to clarify how these variables interact:

—the emergence of scientific consensus on the causes of global environmental change;

—the number of actors responsible for the proximate causes of the global change;

—the nature and global distribution of the harm that might result from inaction;

—the distribution of the burdens of regulation on the consumers, producers, and employees whose behaviors must change;
—the importance of national regulations as a precursor to the emergence of international ones; and
—the need for strong leadership in international forums.

It also suggests that international agreements can be affected by the structures of national political systems, informal international communities, and markets that would be critically affected by agreement.

THE U.S. ENERGY CONSERVATION ACHIEVEMENTS OF 1973-1985

Energy efficiency is probably the most widely accepted strategy for mitigating global warming. The energy shocks of the 1970s led to significant improvements in the energy productivity of Western industrialized economies. The U.S. experience is typical and instructive.

Between 1973 and 1985, the United States reduced its energy intensity—the ratio of energy use to economic output—by 25 percent.[9] Other industrialized capitalist countries made similar achievements, reducing energy intensity, usually from much lower initial levels, by an average of 21 percent during that period (International Energy Agency, 1987). The change was a sharp contrast to the record of the previous two decades and to most of the twentieth century. Between 1953 and 1973, U.S. energy intensity was almost steady, decreasing at an average of 0.1 percent per year; at only two earlier periods in the century, 1918-1926 and 1948-1953, did energy intensity decrease at a rate above 2 percent per year (Schurr, 1984). To the extent that energy intensity can continue to improve in the United States and other countries, energy efficiency can make an enormous contribution to mitigating global warming. This section takes a closer look at how and why the change occurred in the United States and the implications for other countries.

Preexisting Trends

After increasing for 40 years, U.S. energy intensity declined fairly steadily between 1920 and 1953, before stabilizing for 20 years (see Table 4-1). Although the reasons are not well understood, the secular decline in energy intensity since 1920 has been attributed to improved efficiency in energy conversion, a

TABLE 4-1 Average Annual Percentage Rates of Change in Total Output and Energy Intensity in the United States Private Domestic Business Economy, 1899-1981

Period	Total Output		Energy Intensity	
1899-1920	3.4		1.5	
1899-1910		3.9		2.8
1910-1913		4.4		-0.3
1913-1918		2.5		1.6
1918-1920		1.8		-3.3
1920-1953	3.1		-1.2	
1920-1923		5.4		-2.2
1923-1926		3.9		-2.7
1926-1929		2.8		-1.1
1929-1937		-0.4		0.2
1937-1944		6.3		-1.3
1944-1948		0.5		1.2
1948-1953		4.6		-2.9
1953-1973	3.7		-0.1	
1953-1957		2.4		0.3
1957-1960		2.5		0.1
1960-1969		4.5		-0.2
1969-1973		4.0		-0.3
1973-1981	2.3		-2.4	
1973-1979		2.9		-1.9
1979-1981		0.7		-3.9

Note: Subperiods are intervals between business-cycle peak years.
Source: Schurr (1984:Table 3).

shift in the economy away from heavy manufacturing, and technological improvements throughout the economy associated with a shift to more flexible energy sources: oil, gas, and electricity (Schurr, 1984). The decrease in energy intensity with the 1973 oil shock, and again with the 1979 shock, marked a sharp break from the previous 20 years; from 1973-1981, intensity decreased at a rate about 2 1/2 times the average of the previous 53 years.

Uncoordinated Responses to Recent Events

The behavioral change after 1973 was largely due to the oil shocks of 1973 and 1979, which rapidly altered energy prices, changed perceptions of the future price and availability of fossil fuels, and brought about policy changes. Energy users made three effective kinds of responses (U.S. Department of Energy, 1989; Schipper et al., 1990). First, they changed the way they operated energy-using equipment, curtailing heat and travel, and improving management, such as by tighter maintenance of furnaces. Such changes accounted for 10-20 percent of national energy savings achieved in 1986 (compared with the pre-1973 trend; estimates from U.S. Department of Energy, 1989) but are easily reversed when energy prices drop or incomes rise, as they did in the 1980s.

Second, energy users adopted more energy-efficient technology to provide the same service with less energy use, either by retrofitting existing equipment (e.g., insulating buildings, installing reflecting windows) or by replacing existing equipment with more energy-efficient models. These improvements were responsible for 50-60 percent of total energy savings by 1986.

Third, the mix of products and services in the economy changed. Demand fell sharply in energy-intensive industries, such as primary metals, relative to less energy-intensive industries; small cars got an increased share of the automobile market; and commercial airlines improved the match between aircraft size and demand on passenger routes. Together, such shifts accounted for about 20-30 percent of the energy savings achieved in 1986.

Higher real energy prices are generally considered the most important single explanation for these responses (International Energy Agency, 1987; U.S. Department of Energy, 1989). However, price is not the whole story. Although the two energy shocks of the period had very similar price trajectories, the effects on the economic productivity of energy differed markedly after the first two years (see Figure 4-2). For the first two years of each shock, real energy prices increased about 40 percent and energy productivity increased about 5 percent. But over the longer term, the second shock had much more effect than the first. A five-year price increase of about 45 percent in 1973-1978 increased energy productivity 7 percent; a similar increase in 1978-1983 increased energy productivity 18 percent. Moreover, the trend continued through several years of falling real energy prices.

Why the different reactions to the two energy shocks? One explanation is perceptions: it took the second shock to get energy

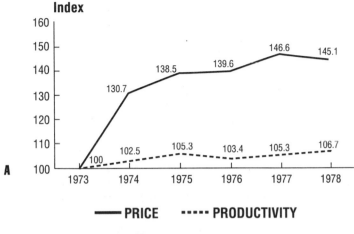

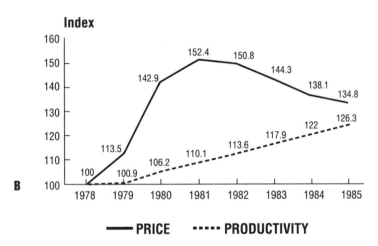

FIGURE 4-2 Changes in indexed real energy prices and energy intensity in the U.S. economy after the energy shocks of 1973 (A) and 1979 (B).

users' attention—to convince them that higher energy prices were here to stay. Another is that the decision environment had changed by 1979 in ways that made it more likely the system would respond to price signals. Government policies to promote energy-efficient technology and provide necessary information were in place by 1979, making it easier for energy users to respond effectively to price; the learning curve for policy implementation had had time to progress; and entrepreneurs were ready to offer energy-efficient technologies and management programs that had not been developed in 1973. Moreover, U.S. energy inefficiency had helped open the door to foreign competition in the automobile, steel, and other industries, with the result that U.S. firms began taking efficiency of all kinds more seriously. Because these explanations reinforce each other, it is difficult to estimate their relative magnitude.

The multiple explanations suggest that the price effect depends on other factors: technological change, policy choices, change in industrial structure, and information processing by energy users. Since these factors can be changed independently of energy prices, it seems likely that with appropriate policies in place, energy intensity might have improved faster than it did, even in the apparently price-responsive 1979-1985 period.

Policy Responses and Implementation

Energy conservation policy in the United States has been predicated on the theory that government should intervene chiefly to correct so-called market imperfections such as the tendency of a supply system based on market prices to produce too little environmental quality (because individual consumers cannot be charged for it) and too little information on energy-efficient technologies and their costs. The government can also intervene to mitigate regulatory and institutional barriers to the functioning of the price system. Following this theory, many U.S. efforts to promote energy efficiency have relied on positive financial incentives (e.g., tax credits, utility rebate programs) and on information. Experience with these efforts shows that the market imperfection theory needs to be expanded to take into account deviations in energy users' behavior from conventional economic rationality.[10] Often, rather than making decisions based on minimization of long-run costs, as theory postulates, energy users act on the basis of nonfinancial values (such as environmental preservation, interest in new technology, or enhancement of social status) or are influenced by information from informal social networks rather than more accurate expert information (see Stern and Aronson, 1984, for a review of evidence). Such processes within individuals and small groups have impeded the effectiveness of conservation programs in the United States, but when they are taken into account, programs became much more effective.

Evaluations of incentive and information programs show that, although they are sometimes very effective at increasing the pace of adoption of available technology, success varies greatly, even between nominally identical programs (Berry, 1990). For instance, home energy rating systems reach between 2 and 100 percent of homes, depending on the market (Vine and Harris, 1988), and utility companies offering exactly the same financial incentive program for home retrofits typically have participation rates that vary by a factor of 10 or more (Stern et al., 1986a).

Success depends on a number of features of implementation. A key is getting the attention of potential participants with appropriate marketing efforts, targeting of audiences, selection of trustworthy sources of information, and other basic principles of communication (Berry, 1990; Ester and Winett, 1982; Stern et al., 1986a; Vine and Harris, 1988; Dennis et al., 1990). Getting people's attention appears to be the main barrier to the success of financial incentive programs for home retrofits, so that, paradoxically, "the stronger the financial incentive, the more the program's success depends on nonfinancial factors" (Stern, 1986:211). Apparently, larger incentives ensure success among those who enter a program but do little to attract participants. Finding the proper intermediary, such as a builder, manufacturer, designer, or lender, can also be critical. Home energy rating systems have been introduced most effectively with the active support of the building and lending industries (Vine and Harris, 1988), and residential conservation programs, especially in low-income areas, have often depended for success on involving highly trusted local organizations, such as churches and housing groups (Stern et al., 1986a). Involving consumers in program design can help fit a program to its audience and locale (Stern and Aronson, 1984).

Thus, conservation policies and programs played a part in the U.S. response to the energy shocks of the 1970s, but they could have had a greater effect with better implementation. Improved policies and implementation, along with higher prices, are among the reasons energy productivity improved faster at the end of the 1973-1985 period than at the beginning. These three factors act in conjunction, however. If, for example, energy prices fall or remain stable, lowering energy users' motivation to change, some policy instruments will become less effective than they were in 1973-1985. The trends of the late 1980s demonstrate this effect (U.S. Department of Energy, 1989).

Implications for Future Climate Change

The technological potential for improvements in energy productivity are huge (National Academy of Sciences, 1991b; National Research Council, 1990a). However, the worldwide prospects for implementing technological changes, and therefore for mitigating the release of greenhouse gases, depends on the behavior of several human systems, including world markets for fossil fuels, national policies for economic and technological development and energy management, global social trends in government and the development of technology, and the behavior of individuals and communities.

The world energy price and supply picture will affect the spread of the Western improvements in energy productivity to other countries. Under conditions like those of the late 1980s, with relatively low energy prices and stable supplies, sharp further improvements in installed energy efficiency are unlikely, even in the Western industrialized countries, without new policy initiatives. The price motive for efficiency is weak, policies that rely on that motive are undermined, and the lowered cost of energy is a spur to economic growth, particularly in energy-intensive sectors. Given continuing population and economic growth, those conditions point to increases in energy use in the wealthy countries, although probably not at pre-1973 rates of increase. A new round of sharp price increases would cut energy use both by reducing economic activity and energy intensity, at least for a period.

The world picture also depends greatly on the development paths of growing economies. Industrialization is energy intensive, enough to have overcome the effects of the 1973-1985 oil shocks in relatively wealthy countries, such as Greece and Portugal, that were still industrializing. Consumers' choices are also important. Where increased income goes into homes and durable possessions, as in Japan, energy productivity is more likely to be higher than where it goes into personal transportation, as in the United States, or into refrigerators or other energy-using appliances, as may become the case in China.

The future of the dissolving socialist bloc countries holds many uncertainties. Many of these countries have highly energy-intense economies and therefore seem to have room for improved energy efficiency given the rise of markets and more democratic control of policy. However, they lack finances to develop technology or implement incentive or information programs and need time to design and implement effective policies for local conditions. Whether development moves in industrial or postindustrial directions is also uncertain. Much room exists for research and for pilot experiments with policy options as ways to reduce the uncertainty.

These and other human systems will determine the extent to which the Western experience with energy efficiency will proceed further or be repeated in other countries. The future will depend on the ways these systems interact in each country and on the ways national and local policies intervene in them.

THE HUMAN CONSEQUENCES OF REGIONAL DROUGHT IN THE SAHEL

Intensification of the greenhouse effect is likely to alter rainfall patterns on a regional scale. As a rule, regions that receive increased rainfall are likely to benefit; decreased rainfall is the more serious concern. The history of the human consequences of severe drought can be instructive about the variety of human consequences of, and responses to, unmitigated climatic change.

The human role in causing drought in the Sahel region of sub-Saharan Africa is a matter of controversy. Throughout the modern history of drought-famine association in the region, there has been a tendency to interpret extreme events as indicators of trends and to attribute the presumed trends to human mismanagement of the local environment. In fact, Sahelian droughts have been recurrent events. The droughts of the 1970s and 1980s were preceded by several others in this century, one of which, in 1910-1915, resulted in intense famine with high mortality. The controversy over the human role in causing Sahelian drought revived with the drought of 1968-1974. The prevailing view was that desertification was an anthropogenic process reflecting deforestation, overgrazing, overfarming, burning, and mismanaged irrigation resulting in salinization of soil and water.

Lack of good data is a major obstacle to understanding the causes of Sahelian drought. Although some evidence supports the orthodox view, some recent research using remote sensing, field measurements, and intensive investigations of small areas has called that view into question. Observable ecological changes are less significant than had been supposed and correlate better with rainfall records than with land management (Mortimore, 1989).

Different Droughts, Different Responses

The consequences of Sahelian droughts in this century have depended on the ability of indigenous systems of livelihood to make adaptations. During the century, these indigenous systems have undergone continual change, first as a result of policies of colonial powers, and later in response to postwar development policies promoting "modernization" and further integration into the global economy. There are competing views of the effects of these century-long trends in political economy on the ability of local populations to withstand drought. In one view, the main results were increased dependency and vulnerability; in the other, vulnerability decreased because of improved availability of medical care, famine relief, and a national infrastructure that allowed for easier migration and food shipments (Kates, 1981).

The three major droughts of the century, in 1910-1915, in 1968-1974, and in the 1980s, have had different effects on the lives and livelihoods of the local populations. The 1910-1915 drought, which was of comparable severity to the drought of the 1970s, appears to have produced greater increases in mortality; its effects on malnutrition and on the social fabric are harder to determine (Kates, 1981). The knowledge base is better for comparing the droughts of the 1970s and 1980s.[11] Local conditions changed between those two periods. Population continued to increase at up to 3 percent annually, forests continued to be cut for fuel and farming, and other forms of resource exploitation probably continued at about the previous rates. Grazing pressure fell, owing to animal mortality but, by the 1980s, cattle holdings had recovered to 60 percent of predrought levels in some areas, and small livestock probably recovered more. On balance, the human demands on the local environment were at least as severe as before the 1968-1974 drought.

The drought of the 1980s was as severe as the previous one. Annual rainfall in 1983-1984 was of the same order as in 1972-1973, and in some areas of the Western Sahel, less. Crop failures and pasture shortages were equally serious. Yet famine did not occur on the same scale, and animal mortality was lower. Possibly food aid was earlier and better in some countries, but in northern Nigeria, where food aid was not a major factor in either period, social distress was noticeably less marked in the 1980s, even in the worst affected areas.

What explains the relatively low human cost of the 1980s drought? It was not the response of the affected governments. Political officials were taken by surprise about equally by both droughts. The people most experienced in surviving failures of agricultural production and managing the environment were those living in the affected areas, but this group had little influence on policy. Of the several political interests concerned with the drought problems, both international and national, the least powerful seems to have been that of the people in the affected areas. Consequently, proposals for new technologies for coping with the drought failed to take indigenous technologies and management systems seriously, and measures to strengthen the poor—for instance by insurance, improved access to resources, alternative job opportunities, and price supports—were rarely considered or given high priority.

A key to drought response appears to have been the role of indigenous forms of land use and response to food shortage. It is possible to distinguish two strategies of land use for areas like the Sahel that face recurrent drought or a long-term threat of declining rainfall. One strategy—maladaptive in the long run—is characterized by deforestation and overcultivation and leads to land degradation, decreases in productivity, and, in the event of drought, short-term collapse. Another—adaptive in the long run—is based on flexible land use, economic diversification, integrated agroforestry-livestock management, and intensive use of wetlands. This pattern tends to generate sustainable, intensive systems and is resilient in the face of drought.

Indigenous strategies of response to acute food shortage apparently enabled the Sahelian populations to survive notwithstanding the tardiness, inadequate scale, and maladministration of most relief programs. These strategies, which relied on economic diversification, such as using labor in urban areas to supplement agricultural income, have evolved in an environment of climatic uncertainty and confer a degree of short-term resiliency. Their future evolution is hard to predict. Continued integration into the world economy may improve roads and other infrastructure, thus enabling diversification; it may also increase pressure for development of cash crops and thus hasten land degradation.

Relationship of Policy to Indigenous Response Systems

The ability of indigenous systems of land use and crisis management to cut the link between drought and famine depends on various factors that sustain the indigenous systems. These include diversity of economic opportunities, absence of war, and appropriate national and international policies on migration. Critical variables include the development of infrastructure and the set of national policies governing access to land, trees, and water. The social distribution of wealth, particularly secure rights of individual or community access to natural resources, determines the extent of human vulnerability to drought. Although some important international actors are coming to perceive these relationships, the political balance is quite different at the national level, where the relevant policies are enacted and enforced. Ruling and military elites, professionals in the civil service, traders (especially in grain), capitalistic farmers, livestock owners, wood fuel exploiters, and small farmers and herders all have separate and distinct interests in the outcome, and most of these interests do not accord high priority to sustainable environmental management or drought preparedness.

Although not enough is known to forecast the consequences of future Sahelian droughts, two alternative scenarios can be imagined. In the doomsday scenario, increasing numbers of people generate cumulative environmental degradation (overcutting of woodland, overcultivation of soils, overgrazing of pastures, and overirrigating and possibly overuse of water), suffer increasing food scarcities as available grain per capita declines, and either starve in huge numbers or migrate in distress to other areas where they become permanently dependent on international relief. In the optimistic scenario, farming systems intensify using an increased labor supply, productivity of the land is raised, sustainable agroforestry-with-livestock systems are extended, and household income sources are diversified and slowly shifted via the market and short-term mobility away from agriculture and toward other economic sectors.

The experience of the 1970s and 1980s suggests that the optimistic scenario is a plausible alternative, given the right policy environment. Its success depends on increased recognition of the potential of indigenous sociocultural systems of land use and household strategies of economic diversification to increase resilience, and on policies that promote resource access and support those local social systems. The consequences of future droughts may also depend on rates of urbanization, growth of the urban informal sector, and capital investment in better favored rural areas. The present policies of governments and international organizations in the Sahel can create conditions that promote or impede the ability of indigenous systems to respond and thus determine the human consequences of future drought.

SEVEN HUMAN SYSTEMS

This section distinguishes seven human systems that may be affected by, and respond to, global change: individual perception, judgment, and action; markets; sociocultural systems; organized action at the subnational level; national policy; international cooperation; and global human systems. It briefly surveys current knowledge and ignorance about the responses of each system and the relationships between them and identifies broad areas in which additional research is needed. It also outlines particular research activities and needs within these areas.

INDIVIDUAL PERCEPTION, JUDGMENT, AND ACTION

The human consequences of global change begin with the individual. Individuals notice the effects of change and either make adjustments or not. Individual behavior is critical in three quite distinct ways: individual judgments and choices mediate responses in all human systems because decision makers begin with inputs from individuals, whether themselves or their advisers. The consequences of global change often depend on the aggregation of the uncoordinated actions of large numbers of individuals. And individual behavior can be organized to influence collective and political responses.

Individual Judgment and Choice

Responses to global changes presuppose assessments of "what is happening, what the possible effects are and how well one likes them" (Fischhoff and Furby, 1983).[12] Scientists, government officials, and other citizens make such assessments when they consider the responses they may make or advocate. Knowledge about human judgment and decision is therefore relevant to understanding responses to global change.

Normative decision principles, such as those of cost-benefit analysis or mathematical decision theory, are limited in their usefulness by the fallibility of the individuals who try to implement them (Fischhoff, 1979); they are even more imperfect for estimating the behavior of people who are not trying (Fischhoff et al., 1982). Past research on human judgment and decision has clarified many differences between decision theory and actual decision making (Kahneman et al., 1982); some of these are reflected in human responses to natural hazards (Saarinen, 1982; Slovic et al., 1974).

Behavioral decision research demonstrates that most people have difficulty comprehending the very low probabilities assigned to environmental disasters (Slovic et al., 1977; Lichtenstein et al., 1978), estimating the probability of natural events that they rarely experience (Slovic et al., 1979), interpreting uncertain knowledge, and making connections between events and their actual causes.

Moreover, it is difficult or impossible to understand unprecedented events and therefore to make wise choices between mitigating them and adapting to them. One result is that lay people frequently perceive environmental hazards differently from specialists (Saarinen, 1982; Fischhoff and Furby, 1983; Gould et al., 1988; Fischhoff, 1989; Kempton, 1991). Little direct knowledge exists, however, on perceptions of climate, climate change, or other aspects of global change (Whyte, 1985; Kempton, 1991; Doble et al., 1990).

Behavioral research also raises questions about expert judgment. Expert analyses, such as represented in general circulation models of climate, inevitably rely on judgment, and judgment becomes more unreliable when the models move into a future different from any past experience. Faith in expert judgments rests on the analysts' success in identifying all the relevant variables and measuring them and their interrelations. Psychological research suggests that people, including technical experts, "have limited ability to recognize the assumptions upon which their judgments are based, appraise the completeness of their problem representations, or assess the limits of their own knowledge. Typically, their inability encourages overconfidence" (Fischhoff et al., 1977; Kahneman et al., 1982). Overconfidence is most likely to affect expert analysts when they lack experience testing their predictions against reality—an inevitable characteristic of predictions about unprecedented events (Fischhoff, 1989). Other kinds of systematic error may also affect experts. For instance, in water resource management and other fields in which average climate parameters are used as a basis for decision, experts seem to exhibit a "stability bias," a tendency to underestimate the likelihood of extreme events (Riebsame, 1987; Morrisette, 1988).

Careful analysts also sometimes overlook or underestimate the likelihood of some possible combinations of events, as they did in a famous assessment of the likelihood of nuclear power plant failure in the 1970s (Nuclear Regulatory Commission, 1978). Little is known about how individuals or groups formulate alternative action plans when faced with a problem, such as responding to a global environmental change. In particular, little is known about what facilitates or impedes creative generation of options, or how vested interest or attachment to the status quo may blind individuals or groups to available options.

Research Needs Research on what and how nonexperts think about particular global environmental problems can help estimate how individuals will respond to new information about the global environment and identify their information needs. This research should address particular beliefs about global change as well as how people evaluate probabilistic and uncertain information and how they combine multiple bits of information from experts, mass media accounts, and personal experience (e.g., with recent weather or air pollution events) to form their judgments about the extent and seriousness of global environmental problems. Such research will require both intensive methods of interaction with informants and survey methods.

Research effort should also be devoted to studying the expert judgment of environmental analysts about global change. This research should address such questions as: Does professional training encourage or discourage particular misperceptions? Does it lead purportedly independent experts to share common preconceptions? How well do the experts understand the limits of their knowledge? Do estimates of the human effects of global change take into account feedbacks among human systems? In analyses of possible responses, what responses are likely to be omitted? To whom do experts turn for analyses of feasibility of responses? What implicit assumptions about human behavior guide the analyses? With preliminary answers to such questions, it is possible to estimate the sensitivity of analyses to variables that affect expert judgment and therefore to make better informed interpretations of these judgments.

Aggregated Individual Responses

The consequences of global environmental change often depend on the aggregated responses of very large numbers of individuals. The example of U.S. energy conservation shows the effect of millions of decisions to buy more fuel-efficient automobiles, reset thermostats, and reinsulate buildings; millions of consumers also drove down sales of aerosol cans when the news got out that they were releasing CFCs harmful to the ozone layer (Roan, 1989). Action to block UV-B radiation from the skin of a billion light-skinned people would similarly take many discrete actions by each of them.

As U.S. energy conservation efforts demonstrate, such individual actions are multiply determined. Financial considerations motivate action, but structural constraints limit action (for instance, not owning the home one would like to insulate); personal attitudes and values increase the likelihood of taking actions that fit the attitudes, subject to the other constraints; specific knowledge about which actions would produce desired effects is helpful, but people often fail to seek it out or mistrust the information available (for reviews of relevant research, see Katzev and Johnson, 1987; Stern, 1986; Stern and Oskamp, 1987). Knowledge has been developed about the conditions under which individuals respond favorably to information (Ester and Winett, 1982; Dennis et al., 1990) and incentives (Stern et al., 1986a) in the context of residential energy conservation; more limited research has been done on other individual actions relevant to global environmental changes.

Research Needs At least three kinds of research should be pursued further to improve understanding of how individual behavior may be significant in response to global change. First is empirical research on the actual responsiveness of behavior to interventions believed to affect it. Energy conservation programs have often produced less than the predicted effects—but as already noted, the responses have been highly variable. For studying possible interventions to mitigate or adapt to global change, pilot studies and controlled evaluation research are particularly important (for a discussion of issues of method in the energy conservation context, see Stern et al., 1987).

Second, new research is warranted to determine the relative contributions and interactions of the various influences on particular individual behaviors implicated in global change (e.g., Black et al., 1985). This research should be interdisciplinary because, in most instances, behavior is jointly determined by technical, economic, psychological, and social variables in ways that are likely to differ as a function of the behavior and the societal context.

Third, research should be conducted to build an improved interface between behavioral studies of resource use and formal models, which are guided mainly by economic assumptions. Empirical analysis of the behavioral processes underlying descriptive categories such as price elasticity, implicit discount rate, and response lag is likely to add to understanding of human responses to price stimuli and government intervention, and also to encourage needed dialogue between economically and psychologically oriented analysts of consumer behavior (Stern, 1984, 1986).

Individuals as Social and Political Actors

Individuals, appropriately mobilized, can be powerful actors at the community and national levels. Individual perception and judgment determines support for social movements, such as the environmental movement, that affect human response by linking individuals to the concerted actions of government and industry. Those actions, in turn, influence individual behavior both directly and through their effects on markets. Individual reactions, in the aggregate, determine the public acceptability of policy alternatives being considered for response. And secular changes in individual attitudes and values, such as about the importance of material goods to human well-being, may have great effects on the long-term response to global change.

Past research has investigated the correlates of environmental concern and related attitudes (e.g., Borden and Francis, 1978; Van Liere and Dunlap, 1980; Weigel, 1977) and tracked the rise of postmaterialist values in the United States and other Western democracies (Inglehart, 1990). Such attitudes have been strong and persistent in many countries since the 1970s. Other research has been devoted to the rise of the environmental movement and to its objectives and tactics (see below).

Research Needs There are important gaps in the literature. New research should carefully assess alternative hypotheses about the links between individuals' values and attitudes and their representation in the activities of environmental movement groups and other institutions involved in response to global change. For instance, the view that environmental organizations reflect widespread attitudes should be tested in the global context against other views, for instance that social movement activists act as entrepreneurs, with their own interests separate from those of the public they claim to represent (e.g., Touraine et al., 1983; Rohrschneider, 1990).

Future research should also address the bases of environmental concern. Such concern may derive from a new way of thinking about the relationships of humanity to the planet (e.g., Dunlap and Van Liere, 1978) or from concern about harm done to people, such as those indirectly affected by market transactions and those yet unborn (Dunlap and Van Liere, 1977; Heberlein, 1977; Stern et al., 1986b). Outside the U.S. context, yet other bases of concern may predominate. For instance, in several Soviet republics, the environmental movement of the late 1980s expressed demands for autonomy by smaller nationality groups against the dominant Russians. On another dimension, environmental concern may derive from personal experience or secondhand accounts in the mass media. The source of concern may determine the conditions under which people become aroused about a global change or receptive to policies that take meaningful action but require additional costs. The determinants of concern are likely to vary with the environmental problem, the country, and characteristics of the individual, so the research should be comparative between countries and environmental problems of different kinds.

MARKETS

One of the most likely consequences of global change will be effects on the prices of important commodities and factors of economic production in local and world markets. As a result, uncoordinated human responses will be affected greatly by markets. According to economic theory, producers and consumers respond to changing relative incomes, prices, and external constraints, so that, if the market signals are allowed to reach individuals and market prices include all the social costs and benefits of individual actions, the responses will be relatively rapid and efficient.

Markets allow for many forms of uncoordinated adjustment, as the example of climate change illustrates. People may rapidly alter patterns of consumption (e.g., substitution of water skiing for snow skiing) and production (e.g., relying on snowmaking equipment rather than natural snowfall). Over the longer run, societies may respond, in the case of unfavorable climatic developments, with the migration of capital and labor to areas of more hospitable climates. Structures tend to retreat from the advancing sea, people tend to migrate from unpleasant climates, and agricultural, sylvan, and industrial capital tend to migrate away from lands that lose their comparative advantage. In addition, technology may change, particularly in climate-sensitive sectors such as agriculture and building.

However, the conditions that economic theory specifies for efficient adjustment are not generally met in the case of the global environment (Baumol and Oates, 1988). In three important respects, existing markets do not provide the right signals (in the form of prices and incomes) of social scarcities and values. And in addition, as already noted, the participants in markets do not always behave as strict rules of economic rationality predict.

Environmental externalities of economic activity, that is, effects experienced by those not directly involved in economic transactions, are not priced in markets today. Someone who emits a ton of carbon into the atmosphere may produce great damage to the future climate but does not pay for the damage: effects that have no price may be treated as if they have no value. Similar problems arise with the externalities of deforestation, CFC emissions, and other environmental problems. Economic theory recognizes that when there are significant externalities, uncoordinated responses will be inappropriate because the market does not transmit the right signals. An additional problem concerns making trade-offs when each response option produces different externalities (Fischhoff et al., 1981; Bentkover et al., 1985; Mitchell and Carson, 1988; Fischhoff, 1991).

The market mechanism is overridden at times, either by political systems (such as when countries set the prices of oil or coal well below or above world market levels); or because custom and tradition determine property rights in a way that precludes the emergence of markets, as in the case of water allocation in the western United States. In such cases, individuals are either not faced with prices at all or are faced with prices unrepresentative of true social scarcities, and their uncoordinated behavior will not achieve the rapid and efficient adjustments characteristic of free markets.

Discount rates in markets, such as interest rates, reflect a social time preference for the present over the future that does not correspond to social valuation of the distant future reflected in concern about problems of global change (Lind, 1986). For events a century in the future, a discount rate that is, say, 3 percent per annum higher than true social preference implies that the future events are valued at only one-twentieth (that is, 1.03^{-100}) of their appropriate value. Market interest rates may be too high to reflect this generation's concerns about the future of the environment; vigorous debate exists about whether the concept of discounting is even moral when human life is at stake (MacLean, 1990). Uncoordinated decisions following such a discount rate undervalue future threats and opportunities.

Economic theory suggests prescriptions for government action when market signals do not correspond to social values. The goal usually considered most important is to get the environmental impacts reliably translated into the price and income signals that will induce private adaptation. But it is difficult to arrive at the "correct" prices because so many of the impacts of global change are unknown or uncertain and because the appropriate values of future events are unlikely to be the same from all generational vantage points and resource endowments (Lind, 1986; Pearce and Turner, 1990).

Economists have suggested some approaches to the problem of developing well-functioning markets to guide responses to global change (for some examples, see Pearce and Turner, 1990 and Dasgupta and Heal, 1979). Theory suggests that governments intervene with policies that meet at least one of these criteria: (1) they have such long lead times that they must be undertaken now to be effective; (2) they are likely to be economical even in the absence of global change; or (3) the penalty from waiting a decade or two to undertake the policy is extremely high. These criteria suggest four kinds of intervention, which we note here.

Government may encourage quasi-market mechanisms before shortages occur. For example, to ensure that water will be efficiently allocated if climate change affects its availability, governments might introduce general allocational devices, such as auctions, to dispatch water to the highest-value uses. The same approach might be applied to allocate land use near sea coasts and in flood plains and to control pollution by auctioning pollution rights. Governments might also support systems of risk-adjusted insurance for flood plains or hurricanes or international climate insurance. These quasi-market mechanisms have both the advantages and the disadvantages of the market. They make allocations efficiently but tend to undersupply goods needed by those who do not participate effectively in the markets, such as people outside the geographical boundaries of a quasi-market, who may receive polluted air or salinated water.

Government may support research and development on inexpensive and reliable ways of slowing or adapting to global change. Research on adaptation is undersupplied by markets because inventors cannot capture the full fruits of their inventions. Research on mitigation technologies that will slow global changes are even more seriously undersupplied in markets, because not only can inventors not collect the fruits of their efforts, but also the fruits, such as preservation of climate, are unpriced or underpriced in the market.

International agreements may provide for international adaptation strategies, such as improved international markets, which allow migration of labor and capital over a greater geographical range than national markets.

Governments may promote needed knowledge and collect and distribute data about global change, to enable rational response. It is difficult for people to mitigate or adapt if they do not understand what is happening or the costs of the available responses and of inaction; costs of adaptation will be reduced to the extent that managers, diplomats, and voters are well informed about well-established scientific results.

Research Needs Although the above market-oriented response strategies are strongly supported by economic theory, knowledge is weak about how they may be effectively implemented. Three lines of research into markets can add to understanding of the available response strategies.

First, empirical studies are needed of the implementation of quasi-market mechanisms for adaptation to global change, to determine how particular mechanisms work in particular social and political systems. For instance, systems for auctioning emission rights can be made infeasible by political opposition, subverted by fraud, undermined by political decisions, or otherwise altered from their theoretically pure operation (Tietenberg, 1985, explains the principle in the case of local air pollution; application to global change would be more difficult). Retrospective and prospective studies of the operation of such mechanisms can illuminate the problems that arise in implementation and assess the actual, as opposed to theoretical, effects of such mechanisms on equity and efficiency. Such assessments should compare quasi-market mechanisms to available regulatory mechanisms, as each actually operates (see the section below on national policy).

Second, studies of the valuation of global environmental externalities are critically important to address several key questions. For instance: To what extent can knowledge or technology be substituted for the outputs of environmental systems, thus making those outputs less indispensable? Is such substitution desirable? How can the "services" produced by the natural environment be included in economic accounting systems, such as national income accounts? How can the producers and recipients of externalities arrive at a common valuation if one side is disadvantaged in financial resources, and therefore in the ability to participate in markets or quasi-markets? How do people value, and make trade-offs between, different kinds of externalities? How do different actors value the effects of human interventions in the environment and make trade-offs between effects? (Some of these questions are addressed in work by Mitchell and Carson, 1988, and Nordhaus, 1990.)

Third, studies of social discount rates are needed, especially to estimate preferences concerning the future environment so they can be included in evaluations of global environmental change (e.g., Lind et al., 1986). Many believe that market discount rates are too high to accurately represent the social value of the future environment, although this value is unknown.

SOCIOCULTURAL SYSTEMS

Between the uncoordinated activities of individuals and the formally organized activities of governments and international organizations lie the oldest forms of social organization: families, clans, tribes, and other social units held together by such bonds as solidarity, obligation, duty, and love. These sociocultural systems have undergone considerable change throughout human history, yet informal groups connected by these bonds still exist and the bonds still influence behavior independently of governments and markets. Sociocultural systems are important in terms of global change in two ways. Some long-lived social units, whose survival may be threatened by global change, have developed ways of interacting with their environments that may be adaptable by others as strategies for response. Also, informal social bonds can have important effects on individual and community responses to global change and on the implementation of organized policy responses.

Indigenous Sociocultural Systems of Adaptation to Environment

Indigenous peoples that were not tightly integrated into world markets have developed technological and social adaptations that often maintain their subsistence in reasonable balance with the local environment. The adaptations of Sahelian peoples to an environmental regime of recurrent drought is one example. A parallel example can be found in the indigenous economic systems on the Amazon, which for at least 500 years have used the ecosystem's material in ways that do not threaten its long-term productivity (Hecht and Cockburn, 1989). The Amazon's indigenous people are a major repository of practical environmental knowledge about sustainable use of resources (Moran, 1990; Posey, 1983). Slash-and-burn cultivation with adequate fallow periods allows for the recovery of vegetation in tropic moist forests (Uhl et al., 1989), attracts game animals to crops (Linares, 1976; Balée and Gély, 1989), and provides a well-balanced, varied diet (Baksh, n.d.). Local agroforestry systems, which combine "the production of crops including tree crops, forest plants and/or animals simultaneously or sequentially on the same unit of land" (King and Chandler, 1978), mimic tropical ecosystems, protecting the soil from leaching and erosion while replicating the natural succession of plant growth over a period of years, and are a model for modern systems of agroforestry. Some such systems can give per hectare yields over five years roughly 200 percent higher than systems established by colonists and 175 times that of livestock (Hecht, 1989a:173).

Agricultural systems based on indigenous models can be profitable in a market economy. Japanese colonist smallholders in the Amazon have created complex systems that prevent soil degradation and tolerate soil acidity and aluminum toxicity better than annual crops. These systems involve polycultures of mixed perennial and annual crops that are transformed, over time, into polycultures of mixed perennials. Commercial quantities of black pepper, cacao, passion fruit, rubber, papaya, eggs, and pumpkins and other vegetables are produced (Subler and Uhl, 1990). Into this sustainable, intensive agroforestry system, the Japanese farmers often incorporate fish culture and chicken and pig production and use waste or refuse from one operation as inputs to other operations (Uhl et al., 1989).

The knowledge about environmental adaptation resident in indigenous social groups depends, of course, on the survival of these groups. Development strategies that destroy the forests can undermine the ability to mitigate or respond to global change by threatening local sociocultural systems based on sustainable, noninvasive strategies of using the land. In the Amazon, the newly expanding, extensive land uses are not compatible with indigenous Indian systems of gathering, long-fallow cultivation, fishing, and hunting and also threaten the subsistence of some 2 million small-scale extractors who collect rubber, nuts, resins, palm products, and medicines while practicing small-scale farming and foraging. Current issues in the Brazilian policy debate that will affect the viability of indigenous groups include the implementation of reserves on which these groups collectively determine resource exploitation (Hecht and Cockburn, 1989), institutions governing the enclosure of public land for unrestricted private uses, and various types of park or biosphere areas with protected wilderness and some degree of zoned multiple use (Poole, 1989:43).

Indigenous sociocultural systems that have adapted to highly variable environments may offer lessons for improving the robustness of social systems to environmental changes outside of past experience. The adaptation in the Sahel points to the importance of diversified sources of cash and subsistence in allowing local groups to adapt to environmental change with limited human cost. An instructive counterexample may be the American Great Plains, where a new generation of settlers between the 1890s and 1920s developed an agricultural system poorly adapted to the area's variable rainfall patterns. The limited adaptability became obvious in the Dust Bowl period of the 1930s. The results included large-scale out-migration and the development of a national system of governmental supports for regional agriculture that encouraged the remaining farmers to further expand their use of limited water supplies. Some analysts believe these changes brought the farmers' adaptability without continued outside assistance into even more serious question (Worster, 1989). Other recent research, however, argues that the serious drought of the 1950s did not have devastating effects and suggests that a recurrence of the climate of the 1930s in the Great Plains would have little effect on the region's agriculture (Rosenberg et al., 1990).

Research Needs Research on intensive, sustainable agricultural systems can help identify and evaluate viable alternatives to development strategies that have resulted in deforestation and land degradation in the tropics. Such research can help develop strategies that may provide subsistence and cash for rural populations but that do not afford the high returns to labor and to speculative activities of unrestricted, extensive land use (Moran, 1990).

Research on systems of land use in variable environments can help identify the characteristics of some of these systems that allow them to take environmental change in stride. Such research can identify anticipatory policies that may enable local or regional social systems to withstand the local effects of global environmental changes at low cost, with limited demands on disaster response systems.

Social Bonds and Responses to Environmental Change

Individual behaviors in response to global change are also affected by informal social influences. People imitate individuals they like or respect, follow unwritten norms of interpersonal behavior, and preferentially accept information from sources they trust (Darley and Beniger, 1981; Brown, 1981; Rogers, 1983; Rogers and Kincaid, 1981).

Such influences are significant factors in social response to natural disasters, particularly those that strike quickly and with little warning, such as floods and major storms (White, 1974; White and Haas, 1975; Burton et al., 1978; Riebsame et al., 1986; Whyte, 1986). Studies of community responses to disaster show that family and acquaintance groups and community organizations are often the focus of behavior (Dynes, 1970, 1972), and that spontaneous improvisation at the local level—often by nongovernmental groups—has been a key to effective response (Barton, 1969; Quarantelli and Dynes, 1977). These findings are relevant to global climatic change in that the consequences of such change are likely to include a shift or increase in the incidence of just such natural disasters.

Informal social links are also significant influences on the acceptance of mitigation strategies, such as energy conservation programs aimed at individuals and households (Stern and Aronson, 1984). Adoption of new, energy-efficient technology tends to follow lines of personal acquaintance (Darley and Beniger, 1981), and participation in government energy conservation programs is higher when the program takes advantage of personal acquaintance-ships and local organizations with good face-to-face relations with members of the target group (Stern et al., 1986a).

Research Needs Efforts to develop policy responses in anticipation or response to global change will benefit from knowledge of sociocultural systems of social influence. Research efforts can profitably focus on understanding the social networks, norms, and influence patterns of groups that are highly likely to suffer from anticipated environmental change, so that policies can be designed to work with rather than against these lines of influence. Policy studies should focus on ways to directly involve affected groups, and should compare implementations of the same policies with and without such efforts.

ORGANIZED RESPONSES OUTSIDE GOVERNMENT

Three kinds of social actors other than governments may make significant, organized responses to global change: communities, social movements, and corporations and trade associations. These collective actors form a vital link between behavior at the level of individuals, firms, and households and at the level of institutions and nations.

Communities

A community is more than a shared place of residence. It is also a unit in which people earn their living, engage in political activity, raise their children, and carry out most of their lives. Community responses to the stresses of environmental change occur both in the uncoordinated ways discussed in the previous section and through organized activity. Decades of research on economic development in rural areas suggests that the full impacts of major social changes, including those that may be induced by environmental change, can be understood only by considering the effects of such changes on communities, as well as on individuals and institutions (e.g., Field and Burch, 1988; Machlis and Force, 1988; Machlis et al., 1990).

Communities are likely to respond in different ways to the local impacts of global environmental change. Some communities are sufficiently diverse to provide valuable buffers against hardship as individuals and households share resources. But if all members of the community use the same environment in similar ways, no such buffering is possible. Traditional relationships and patterns of action, tension, and rivalry within a community may help the community through crisis, or may prevent organized action that would help the community cope with or take advantage of local changes. And if local manifestations of global change disrupt traditional patterns of community life, they generate stress and conflict that can become violent.

Of course, the character of community life continues to change in much of the world. With the rapid growth of urban and suburban areas in the developed and especially the developing world, the historical links among home, polity, and economy are greatly weakened. The spheres in which individuals and households act become more disjunct and less well integrated. Global environmental change may increase the pace of this historical trend if it makes rural agricultural life more difficult and thus increases the migration to urban areas, with consequences for the ability of communities, particularly in the Third World, to withstand further environmental change.

Research Needs Research is needed on those characteristics of communities that affect their organized responses to global change. For example, in the United States, the spatial character of suburban communities is a significant barrier to increased use of public transportation. Yet some suburban communities and small towns have been vigorous in their implementation of environmental and energy and water conservation policies (Dietz and Vine, 1982; Berk et al., 1980; Vine, 1981). The response of those communities seems to be greater than would be expected from aggregated simple self-interest or the technical response to changes mandated by policy. The community amplifies individual action, perhaps by creating a sense of identity and trust that overcomes the usual collective goods problem. Especially in the less-developed world, effective community response may depend on the community's access to a variety of resources that can be used to dampen adverse changes in any single resource. In addition, adaptation by individuals and households may be conditioned by the diversity and flexibility of the community, which are in turn affected both by the natural environment and the local political economy, history, and culture. Research is also needed on the conditions controlling the differential effectiveness of environmental and energy programs in different communities.

Social Movements

Environmental movement organizations have been major actors in debating national and even international responses to global change (also see the section below on national policy). The broad awareness that global changes are occurring is in large part due to various national environmental movements drawing attention to the growing body of scientific evidence on the subject.

Most of the national activity of environmental movement organizations is intended to change public policy. How environmental groups influence policy depends on the political context in which they operate, and in particular on the relationship between the movement and political parties. In political systems in which it is difficult to achieve participation via a small party, such as the United States, movements have only loose alliances with political parties. In systems where small parties can play a serious role in influencing policy, the movements either form tight alliances with parties or act as parties in themselves. These structural differences affect movement strategy and have produced some sharp differences in how environmental problems are conceptualized. The ways political structure affects the political impact of the environmental movement on policy have not been studied in enough detail to offer generalizations.

Whatever their relation to political parties, environmental groups usually find themselves in conflict with corporations, trade associations, and often with government officials. Each side brings a different mix of resources to the conflict. In the United States, environmental groups seem to have a high degree of public support and strong legitimacy with other actors in policy debates (Dietz and Rycroft, 1987). Corporations and their representatives have far greater financial and personnel resources, but less public support and less legitimacy within the policy system. Government falls between the two. The difference in resources means that each group will struggle not only over the substance at issue, such as a specific policy, but also over the definition of the problem and the kinds of resources that are legitimate for resolving the problem (Dietz et al., 1989). The difference in resource distribution has typically led industry to favor heavy reliance on scientific analyses and technologically driven policies, and led environmentalists to be more skeptical of those alternatives and inclined to favor source reduction and infrastructure changes.

Modern environmental groups play an important role in shaping public values and consciousness. Indeed, some students of the movement have suggested that its primary goal is to change ways of thinking rather than specific political choices (Cohen, 1985; Eder, 1985; Habermas, 1981; Offe, 1985; Touraine, 1985; Touraine et al., 1983). The rise of "green" ideologies in the United States, Western Europe, and throughout the world seems to reflect changes in consumer preferences and lifestyles that may have important implications for individual, household, and community response to global change (Inglehart, 1990).

Research Needs A number of important questions need to be answered about the role of the environmental movement in responses to global change. How do the strategies pursued by environmental movements in both the developed and less-developed nations influence the character of national policy? What impacts do these influences have on the ability to reach international accords? How does environmentalism interact with scientific research on global change, and what could be done to produce better interactions? How much change in individual ideology is brought about by the environmental movement, and how do these changes affect the behavior of individuals, households, communities, and other actors? What is the likely character and influence of the environmental movements that are emerging in Japan, Eastern Europe, and less developed nations and what role will they have in shaping national and international response?

Corporations and Trade and Industry Associations

Corporations and trade and industry associations are major actors shaping response to global change. Just as the environmental movement translates public concern into political action and in turn shapes public perceptions and actions, corporations and trade associations translate the interests they represent into political positions and also educate those connected to them. As already noted, these groups come to the policy arena with very different resources than environmental groups and, in general, tend to favor different methods for analyzing environmental problems and different strategies for solving them (Dietz and Rycroft, 1987; J.R. Wright, 1990).

Research Needs The relationships of corporations and trade associations to national policy systems, critical for understanding policy response, are discussed in the next section. The internal aspects of these collectivities, however, are little studied. Corporations communicate with each other, and trade associations are influential in shaping the response of corporate members, two processes that shape the policy positions of the business community. Research is badly needed on how corporations and trade associations attempt to communicate internationally about global environmental issues with other groups representing the same industries.

NATIONAL POLICY

Nation-states help determine the consequences of global change through their essential role in international agreements and by national policy decisions that affect the ability to respond at local and individual levels. This section focuses primarily on two issues: differences between nations in their environmental policies and the policy process.

National Differences in Environmental Policy

National environmental policies vary in part because of different public attitudes. People around the world have shown concern with the environment, but the intensity and focus of interest have varied from country to country. Some observers claim that during the early 1970s, environmental issues were much more politically salient in Japan and the United States than they were in Europe; during the 1980s, the reverse has been true (Vogel, 1990). Such variations may be a function of national economic performance, actual environmental quality, or national political cultures. The focus of environmental concern in Japan has been claimed to be on the protection of public health, while in Germany the protection of nature has been accorded much higher priority, with the United States and Great Britain falling somewhere in between (Vogel, 1990). These differences, which may be more or less stable over time, are likely to have important implications for different nations' responses to various kinds of global environmental issues.

Policies also vary because each nation's political system responds to public concerns in its characteristic way. Within democratic nations, many political features vary. Nongovernmental organizations concerned with environmental improvement are not equally well organized in all countries. Citizens of different nations display different propensities to join voluntary organizations concerned with environmental improvement, and these organizations do not have similar access to the policy process everywhere. The United States, with its constitutional system based on the separation of powers, provides nongovernmental organizations with substantial opportunities to shape public policy through access to the courts and the national legislature. By contrast, more centralized political systems, such as France and Japan, severely restrict participation by citizens' groups. Parliamentary systems that have proportional representation, such as in Germany, provide access to the political system by facilitating the formation and representation of political parties committed to environmental improvement (see Parkin, 1989).

Policy systems also vary in the response of major affected interests, particularly those of business. Most environmental problems, domestic as well as global, require substantial changes in what firms produce and how they produce it. To the extent that these changes increase costs, businesses are likely to oppose them and the changes are unlikely to occur. Business resistance can be reduced if new technology enables firms to behave in ways that are environmentally benign without increasing their costs, if consumers develop a "green" consciousness that opens new markets, or if government offers subsidies. As a rule, environmental policies are more likely to be effectively implemented to the extent that investors and managers in some industries and firms believe they can benefit financially.

These issues extend beyond business. Environmental regulations do not simply impose additional burdens on producers; they also affect the relative welfare of consumers, employees, and taxpayers. These burdens may be primarily nonmonetary or monetary, concentrated or dispersed, and relatively visible or invisible, but in all cases they have important political consequences. There is a relevant body of research on how interest groups respond to different kinds of expected burdens and benefits, at least in the United States and a few other countries (Leone, 1987; Meiners and Yandle, 1989).

Environmental policy systems vary in many ways in their approaches to regulation (e.g., Tarlock and Tarak, 1983; Mangun, 1979). Regulations may control emissions at the source, by establishing environmental quality standards, or by establishing exposure standards. Each strategy has various strengths and weaknesses (see Haigh, 1989). Environmental regulation can be coordinated by a single regulatory body or dispersed among a variety of regulatory authorities; relatively centralized in the national government, as in Great Britain, Japan, and France, or administrated primarily by local governments, as in the Federal Republic of Germany. Regulation can be anticipatory, requiring firms to get permission before they can act, as with mandatory environmental impact assessments, or may take place after the fact. And there are different national styles of regulation (Vogel, 1986). The United States has developed an adversarial regulatory style, in which government establishes ambitious and highly specific standards and frequently tries to impose legal penalties for noncompliance. Great Britain, by contrast, uses an approach to regulation characterized by more flexible standards, modest goals, very infrequent use of legal penalties, and restricted participation by the public and environmental groups.

Scientists and scientific evidence play very different roles in different countries' environmental policies. The United States is unusual in providing opportunities for diverse groups of scientists to affect regulatory policies. By contrast, participation by scientists in Europe is more likely to be confined to official channels. The United States is also unusual in having regulatory decisions tied by statute to the outcomes of risk analyses. Thus, it is sometimes easier to have a product or production process banned or restricted in the United States than in most other capitalist nations (see, e.g., Brickman et al., 1985).

Research Needs Most of the sources of variation mentioned apply not only to environmental policies but also to national-level policies in many other areas that can have significant effects downstream. Research is needed to assess the effects of national macroeconomic, fiscal, agricultural, energy, economic development, and science and technology policies on global change and on the ability to respond to global change. These effects are much less well researched than the effects of environmental policy.

Cross-national research comparing the determinants of national environmental policy, focused especially on responses to global change and on the sources of policy differences between countries, is also needed. This research should assess the effects of influences such as public opinion, environmental movement organizations, and various organized interest groups, as well as structural features such as democratic versus nondemocratic politics, market versus centrally planned economies, relative wealth, scientific and technical resource base, and position in the world political-economic system (studies of this type, not focused on responses to global change per se, include Brickman et al., 1985; Jasanoff, 1986; Vogel, 1986; and Jasper, 1990). Such research can help clarify the kinds of policy options that are viable in different countries, which is a factor in reaching and implementing international agreements. In particular, it is important to understand the conditions under which nations enact policies promoting the development of environmentally benign technologies because such development, while it could produce large benefits on a global scale, is often unlikely to come from the private sector because of the difficulty of appropriating profits.

Research should also assess the impact of environmental regulation and alternatives to regulation cross-nationally and across policy questions to clarify how, why, and under what circumstances different regulatory or other strategies work in different policy settings. Such research should proceed despite the lack of clear standards for comparing the effectiveness of the environmental policies of different governments. Every indicator has both strengths and weaknesses. For example, emissions and environmental quality are affected by many factors other than policy, including topography, the nature of industrial production, and the rate and location of economic growth. Likewise, expenditures on abatement by industry are an imperfect measure of the effectiveness of regulation because they may or may not represent a net economic burden. The useful literature on policy compliance and effectiveness is largely confined to a handful of countries and policies (e.g., Bardach and Kagan, 1982; DiMento, 1989).

Finally, research should compare the institutions used in different countries to manage conflict over environmental policy. These institutions are both formal (e.g., legislative and regulatory proceedings and court decisions) and informal (e.g., lobbying, use of publicity in the mass media), and they deal with substantive disagreements, formal procedures, and disagreements about the nature of knowledge about global change and the likely impacts of policy choices. Distinctive national systems of conflict management can be identified and compared; each probably generates characteristic patterns of conflict and characteristic difficulties in decision making.

The Environmental Decision-Making Process

The consequences of global change depend on decisions made in government agencies and other large organizations. Knowledge about the decision process in such organizations is therefore potentially relevant to responses by both governmental and nongovernmental organizations. Specialists on decision processes, a field that makes no sharp distinction between governmental and other complex organizations, typically distinguish analytically among phases of the process, such as understanding the phenomena, identifying viable options, and selecting an alternative.

Government agencies involved in responding to global change rely on information from experts to gain understanding, but they must make it useful to their leaders, who are almost always non-experts, and must interpret the conflicts between, and uncertainty within, expert judgments. There is a general body of literature on the ways government agencies and other large organizations acquire and process expert knowledge (e.g., Lindblom and Cohen, 1979; Weiss and Bucuvalas, 1980) and on the inherent problems of informing nonexpert decision makers about uncertain and disputed scientific knowledge (National Research Council, 1989b).

Organizations can generally identify a large number of options, but they tend to funnel information to narrow the universe of issues or action alternatives presented to leaders (March and Olsen, 1989). Similarly, not all options known to a society reach its legislative agendas (e.g., Kingdon, 1984). Among the factors involved in getting environmental issues on political agendas are mass media coverage of disastrous or telegenic events and threats of dread consequences such as cancer, danger to children and future generations, the characteristics that increase perceived seriousness of risks among most citizens (Mazur, 1981; Sandman et al., 1987; Rosenbaum, 1991). Government action on environmental hazards is typically driven by crises, with major events evoking bursts of legislation (May, 1985; National Research Council, 1987). It is less clear, however, how particular response options get on the agendas of government agencies or other organizations.

Decisions within government agencies and other large organizations are affected by standard operating procedures, preassigned divisions of labor, accounting systems, organizational cultures, bureaucratic politics, organizational hierarchy, bargaining and negotiation processes, leadership practices, and the control of information by constituent individuals and subunits with goals only partly coincident with those of the organization as a whole (Seidman and Gilmour, 1986; March and Olsen, 1989). Decisions are influenced by relationships between organizations, for example, in international environmental agreements, interagency negotiations, lobbying coalitions, and even large industrial firms that must weigh the positions of their marketing, manufacturing, engineering, and legal departments in deciding whether to change to a more environmentally benign manufacturing process. Decisions are also affected by the structure of institutions—the systems of rights and rules that constrain the actions of individual parties. Examples include the effects of such institutions as markets for land and energy, land tenure systems, the law of property rights and torts, representative government, and international regimes (discussed in the next section).

Research Needs The organizational decision-making perspective points to a number of areas in which the general concepts in the field might be usefully applied to organizational actions affecting response to global change. For instance, informative studies could be done on how organizational understanding of environmental issues develops; how intraorganizational factors affect the responses of corporations, government agencies, and national political systems to global change; and how bargaining, rivalries, informal norms, and other processes of influence between organizations affect organizational responses to global change. An area of more pointed interest is the comparative study of environmental decisions in different institutional contexts. To gain understanding of the consequences of global change, it is important to understand the effects of different systems of land tenure on deforestation, of different national regulatory systems on the control of atmospheric pollutants, and of different systems of property rights in subsurface resources on policies to limit extraction of fossil fuels.

INTERNATIONAL COOPERATION

Sustained international cooperation is one essential element in the overall human response to global environmental changes. It is essential because efforts to cope with some large-scale environmental changes such as ozone depletion and global warming seem doomed to fail if some of the major national actors do not cooperate. Recent agreements among the advanced industrial countries to phase out the use of CFCs cannot solve the problem of ozone depletion unless some way is devised to persuade China, India, and other developing countries to use substitutes for CFCs in their rapidly increasing production and consumption of refrigerants. The global warming problem is even more complex. Not only is there a need for cooperation between the advanced industrialized states and the major fossil fuel-using states of the developing world, but there is also the problem of controlling other sources of greenhouse gases. These sources are as diverse and widespread as methane-releasing agricultural activity in south Asian rice paddies and North American feedlots and carbon releases from cutting tropical forests in Zaire and Brazil.

Some environmental problems call for international action because activities in one country produce spillover effects or externalities affecting other countries. An example is the emission of airborne pollutants in the eastern United States and Eastern Europe. International cooperation is needed to articulate and apply liability rules or to allow the countries affected by spillover effects to compensate those responsible for the offensive emissions for terminating or redirecting their activities.

Today's concerns with international arrangements focus mainly on mitigating global environmental changes rather than adjusting to them. In the future, however, as global changes become realities, there will be more calls for international cooperation to adjust to the impacts, for instance, by developing buffer stocks of food crops or mechanisms to handle flows of environmental refugees.

International cooperation poses difficult problems, even when all the parties stand to gain from the right agreement. One of the most robust theoretical findings of the social sciences is that rational actors engaging in interactive decision making in the absence of effective rules or social conventions often fail to realize feasible joint gains, sometimes ending up with outcomes that are destructive for all concerned (Olson, 1965; Hardin, 1982). The conditions of international society make the problem more complicated than it is in other situations. The issues are seldom well defined at the start, so that preliminary negotiations may be needed to define them. When unanimity is required, some states can hold the agreement hostage to better terms for themselves. Each country is complex, and bargaining within countries can make international agreements especially difficult (Putnam, 1988). And the agreement can take second place to more immediate issues in any of the countries involved.

Most observers now believe that the key to solving these collective-action problems is in the creation of international regimes, or more broadly, international institutions (Krasner, 1983; Young, 1989a). Regimes are interlocking sets of rights and rules that govern interactions among their members with regard to particular areas of action. Although most of the research on international regimes concerns economic regimes, interest is mounting rapidly in the study of environmental regimes, particularly the developing regime for the protection of the stratospheric ozone layer (Benedick, 1991), but also other, more geographically limited, international environmental regimes (e.g., Sand, 1990a; Haas, 1990).

The ozone regime exemplifies one model of regime formation, in which a framework convention is followed by a series of substantive protocols in quick succession. Another model sets out substantive provisions in more or less complete form in initial agreements. Cases in point include the 1946 International Convention for the Regulation of Whaling and the 1973 Convention on International Trade in Endangered Species of Wild Fauna and Flora (Lyster, 1985). Additional study is warranted to determine the circumstances under which one or the other of these models is more appropriate.

Most of the research on environmental regimes has so far emphasized regime formation, particularly the determinants of success or failure in forming regimes and the timing and content of successfully formed regimes. This work has highlighted five sets of explanatory variables. One stream of analysis emphasizes structural aspects of the relationships involved in regime formation, such as the number of participants, the extent to which interaction is ongoing, and the nature of the mixed incentives to cooperate and compete (Oye, 1986). Another stream focuses on the role of power relationships, such as the presence of a hegemonic power, that is, an actor possessing a preponderance of material resources (Keohane, 1984:Chap. 3). A third stream emphasizes factors likely to impede or facilitate the negotiation process, such as the extent to which negotiations lend themselves to "integrative bargaining," the thickness of the "veil of uncertainty," the impact of exogenous crises, and the role of leadership (Young, 1989b). A fourth stream emphasizes cognitive variables, such as the role of widely shared ideas (Cox, 1983) or an "epistemic community," that is, an international group of officials and scientists who share a set of causal beliefs and a set of preferences for action (Haas, 1990). A final stream of research stresses the importance of the international context in providing windows of opportunity for agreements that are blocked at other times by resistances in one country or another.

Research Needs Knowledge is limited on several aspects of international agreement that are particularly relevant to problems of response to global change. One is the effectiveness of institutional arrangements, that is, the factors determining how strongly a regime affects the behavior of those subject to its provisions. Effectiveness is partly a function of implementation which, as at the national level, often leads to outcomes quite different from what a reading of the initial agreement would lead one to expect (Pressman and Wildavsky, 1984). It also depends on the degree to which arrangements are structured so that those subject to the regime comply voluntarily and do not have to be continually monitored and coerced. Finally, it depends on the ability of a regime to persist even after the constellation of interests that gave rise to it has changed or disappeared (Krasner, 1989).

Another area for new research concerns preparatory negotiations, aimed at reaching a common conceptualization of environmental problems. Many international issues that require cooperation are not ripe for negotiation because the issues have not yet been defined in a way suitable for bargaining (e.g., Stein, 1989; Saunders, 1989). This certainly seems to be the case for complex environmental issues, such as would be raised in drafting a comprehensive law of the atmosphere on the model of the law of the sea. National representatives would need first to identify packages of policies they might use to comply and assess the costs of those packages in terms of their interests. The process would be much more complex than establishing limited regimes to deal with ozone depletion or acid rain or establishing a series of regional regimes combined with agreements between regional groups.

A third area concerns the problems of regime formation when the participants are deeply divided. Many global environmental problems involve north-south confrontations in which the wealthy, industrialized states want to limit environmental changes but developing countries see limits as threats to their development. Examples include conflict between the desire to limit carbon dioxide emissions and energy needs in China and India, and between the desire to protect global biodiversity and plans for the use of forests in Brazil and Indonesia. Much needs to be learned, for example, about the bargaining power of apparently weak players, like China, which can issue credible threats to step up their use of coal or CFCs unless others make it worthwhile for them to desist.

More knowledge is also needed about the role of nonstate actors, such as intergovernmental organizations, environmental movement organizations, and transnational corporations, in the creation and operation of environmental regimes. The involvement of such nonstate actors heralds the emergence of a more complicated international society in which states remain important but share influence with several other types of actors. This change may require more sophisticated conceptualizations of international interactions.

Finally, there is need for better understanding of the relationships between institutions (sets of rights and rules) and organizations (material entities with offices, staffs, budgets, and legal responsibility) (Young, 1989a, b). Organizations, such as the United Nations Environment Programme, have sometimes been important players in regime formation; they are sometimes necessary to manage regimes, although implementation of key rules is sometimes delegated to the member governments. Given the costs of operating international organizations, it is important to have a better understanding of the conditions under which they are necessary, or more effective than alternatives.

The above research agenda is relevant not only to the practical problems of responding to global change, but also to some basic issues in social science. The gaps in knowledge about international environmental regimes are also gaps in the broader literatures on social institutions and collective action. This global change research agenda would therefore be a direct and timely contribution to political science.

GLOBAL SOCIAL CHANGE

As we note at the opening of this chapter, the consequences of global environmental change depend on the future shape of human society. A number of ongoing changes in human systems, operating systemically or cumulatively at the global level, are shaping the societies that will feel the effects of global environmental change. Although global social changes are numerous, to our knowledge, a thoughtful typology of them has not been developed. As an impetus to further analysis and research, we note several examples of global social changes that may affect the driving or mitigating forces of global environmental change or the ability of human systems to respond to such change.

1. *Population Distribution and Size* The urban population of the world continues to increase both in total and in percentage terms, in both the developed and developing countries (Berry, 1991; Smith and London, 1990). Urbanization, by increasing spatial concentration, may increase vulnerability to natural hazards, concentrated pollutant emissions, and globally systemic changes such as sea-level rise. Urban bias in developing countries may also skew national priorities away from rural resource and environmental problems (Lipton, 1977). However, urbanization may decrease vulnerability by affording economies of scale in resource use and environmental protection, allowing rural households to diversify their sources of income, decreasing population growth rates, and increasing concern with environmental amenities. Some of the key research questions concern the conditions under which urbanization affects demand for resources implicated in global change, vulnerability to environmental disasters, and the robustness of rural communities in the face of environmental change. Equally relevant are concerns of population size. Increasing human population is likely to place added pressure on political and economic systems to contain conflicts likely to arise over increasingly scarce resources (see, e.g., Homer-Dixon, 1990).

2. *Market Growth and Economic Development* The spatial reach and dominance of market forces have been widening as a world system of trade penetrates even into countries that have had central planning and command economies and into the remotest regions. The effects on the human driving forces of global change and on the ability to respond are not obvious. Expansion of the market replaces state-sponsored resource waste with an invisible-hand means for checking inefficient and degrading uses of the environment. However, ceding control to the market can also lessen the ability of the state or community to manage environmental problems that are driven by the search for profits. At the local level, sustainable practices associated with a subsistence or mixed economy may be abandoned for unsustainable profit-oriented ones (Bates, 1980; Jodha and Mascarenhas, 1985; Redclift, 1987). The increased wealth that is the usual (though not always realized) goal of a shift toward free-market policies generally increases the ability to respond to threatening changes;

it may also raise the standard of environmental quality expected by the population.

3. *Socioeconomic Marginalization*   Some observers hypothesize that the global spread of capitalism has forced certain individuals, groups, and countries into a position of diminishing control over needed resources and reduced options for survival and for responding to global change.   Indigenous sociocultural systems of social security are believed to be crumbling, with new capitalist economies doing little to replace the lost safety nets. Economically marginalized individuals and groups sometimes degrade the environment for subsistence and lack the resources to respond effectively to natural or human-induced damage.   Marginalization and impoverishment of nations can have the same consequences for national policies and actions (Hewitt, 1983; Sen, 1981; Watts, 1987).

4. *Geopolitical Shifts*   The trend in 1989-1991 of declining tensions between East and West may facilitate human response to global environmental change through reallocating funds from military uses, lowering the potential for widespread nuclear and/ or chemical warfare, redefining national security to consider environmental as well as military and ideological threats (Brown, 1982; Mathews, 1989; Bush and Gorbachev, 1990), and building trust between powerful nations that will lead to cooperation instead of conflict.   At the same time, however, north-south tensions may be increasing with the disparity of wealth between the developed and developing worlds.   Such increased tension will make future international cooperative action more difficult and may lead to direct conflict (Agarwal, 1990; Carroll, 1983).   The net effect of such geopolitical shifts is very hard to predict.

5. *International Information/Communication Networks*   A global explosion of information and communication technology has uncertain implications for response to global change.   It may facilitate societal response by making it easier for scientists and policy makers around the world to cooperate and share information, disseminate it to the public, and marshal worldwide pressure for response (Cleveland, 1990; Miles et al., 1988; Mowlana and Wilson, 1990; K. Wright, 1990).   Examples include international reaction to satellite photographs of daily burning in the Amazon forests and the response of the Soviet peoples to news of the desiccation of the Aral Sea.   However, the network may also amplify misinformation or create barriers to response by spreading the word that some nations may gain from environmental change.

6. *Democratization*   As of mid-1991, there appears to be a worldwide trend toward increasing decision-making power of the

citizenry in nation-states. Increasing democratization may influence human response by providing more power to people being affected by environmental change, but it may also give more access and power to those whose interests would be harmed by measures for environmental management and protection. Democratization may also slow responses, compared with what might be achieved in an authoritarian regime by simple decision by the leadership (Kaplan, 1989; Muller, 1988; Roberts, 1990; Stephens, 1989). The net effects on response to global change are likely to depend on conditions in particular countries.

7. *Scientific/Technological Expansion* Exponential growth in scientific and technological knowledge both drives environmental change and increases the capacity to respond to it. It increases the ability to detect and understand threatening global environmental changes (e.g., the ozone hole) and provides alternatives to destructive products and practices (e.g., substitutes for CFCs) (AMBIO, 1989; Bacard, 1989; United Nations, 1989), but it may also create new global environmental problems (Kasprzyk, 1989; Russell, 1987). And new technologies may create major changes in the structure of human society, as in the case of CFC refrigeration technology or the periodic emergence of new energy sources to replace old ones as the basis of industry (Ausubel and Sladovich, 1989). In such instances, the implications for the global environment may remain uncertain for a long period.

8. *Resurgence of Cultural Identity* Many analysts perceive a worldwide resurgence of cultural identity or differentiation in recent decades: a deeply held attachment to groups (e.g., ethnic, religious, tribal, states) and the associated movements by these groups for autonomy of expression and decision (see Nash, 1989). Examples include the resurgence of ethnic nationalism in the Soviet Republics and the overt hostility, especially in Islamic countries, to the cultural invasion of Western values. The impact on response to global change is most likely to be felt when global changes or possible responses to them are perceived as threats to the values or livelihood of a particular group or when response requires cooperation between groups already in conflict.

The social changes mentioned appear to be ongoing trends, yet their future direction is, of course, uncertain. Equally uncertain are the effects of any trends in global human systems on the human ability to respond to global change. Plausible arguments can usually be made on both sides: a global social change may make resource use either more or less extensive and effective human response either easier or harder to accomplish. The open questions point to many research opportunities for social scientists who have studied changes in these human systems and who would now consider their implications for human responses to global change.

CONCLUSIONS

This chapter examines the range of human consequences of global change and identifies specific areas in which new research can make important contributions to understanding. Where we identify research needs, priorities among studies should be set according to the criteria noted in Chapter 2. We focus here on four general principles derived from this analysis that deserve special emphasis because they are fundamental, underappreciated, and point to critical directions for research.

THE KNOWLEDGE BASE FOR HUMAN RESPONSES IS INHERENTLY VALUE LADEN

We have identified the key link from environmental change to its human consequences as proximate effects on what humans value. Of course, what humans value depends on the humans. The wealthy tend to have different value priorities from the poor, national leaders from voters, business executives from laborers, miners from herders, and so forth. Yet what humans value is precisely what defines the consequences of global change and drives human responses. Different individuals and human groups will often disagree about what environmental changes are worthy of response.

Research Needs First, it is necessary to disaggregate the consequences of global change by analyzing the distribution of impacts of particular global changes on the things that different groups of people value. Such knowledge is necessary input to policy debates, even though it is not sufficient to facilitate social choices. Even with perfect knowledge of the effects of each conceivable alternative on each group affected, conflicts of value and interest will remain. Better knowledge of the impacts may even precipitate conflict by making latent conflicts more obvious.

Second, it is important to develop better ways of making the available knowledge about outcomes more accessible and understandable to nonspecialists. The body of knowledge about the design of messages about environmental risks and benefits can be brought to bear (National Research Council, 1989b; Mileti and Fitzpatrick, 1991). Better messages are also necessary but insufficient to facilitate social choices. They inform but do nothing to alter the differences in values and interests that produce conflict.

Third, it may help to understand the process of value judgment better. Several systematic methods have been used to assess the value people place on outcomes that may be affected by environmental change or responses to it, and to help individuals confront the value trade-offs that policy choices often pose (e.g., Keeney and Raiffa, 1976; Mitchell and Carson, 1988). These methods of systematizing the valuation process can be applied to the valuation of the consequences of global change under different response regimes; such studies will advance understanding of valuation and may also help individuals and social groups choose their responses.

The most critical practical need is probably for effective means of managing the conflicts of value and interest that attend choices about global change. Human systems at every level of organization will have to develop systems of conflict management and, to the extent that different human groups (e.g., countries) need to respond in a coordinated way, their systems will also have to be compatible. These practical needs raise numerous research questions for the global change research agenda. In the discussion of conflict, we noted several bodies of relevant theory and knowledge that could be usefully applied to the study of conflict over responses to global environmental change. Methods of conflict management developed for other conflicts might be tried experimentally and monitored in efforts at global change-related conflict resolution. And experiments should be conducted with institutional means for making technological knowledge useful to nonexperts in a context of controversy—for instance, systems that enlist representatives of interested groups in the process (National Research Council, 1989b) or that harness the controversy to provide a range of perspectives as an aid to understanding (Stern, 1991).

HUMAN RESPONSES MUST BE ASSESSED AGAINST A CHANGING BASELINE

The human consequences of an environmental change depend on when it happens and on the state of the affected human groups at that time. Global changes in the future may or may not have more serious effects than if they happened now. For instance, if recent trends continue, future societies will be wealthier, more flexible, and more able to take global changes in stride than present ones. However, the more committed human societies become to present technologies that produce global change, the harder it will be to give them up if that becomes necessary.

Research Needs First, to understand the human consequences of global change, it is important to improve the ability to project social change. Existing methods range from simple extrapolation to more complex procedures for building scenarios. But scenario building is more art than science. Therefore, as an initial approach, it is useful to test projected environmental futures against various projected human futures to see how sensitive the human consequences of global change are to variations in the social future. In the longer run, it is much preferable to improve understanding of the relationships that drive social change. This is a long-term project in social science, on which much theoretical work is needed. We return to this theme in Chapter 5. Research on the human dimensions of global change may help give impetus to that project.

Second, the extreme difficulty of predicting the long-term social future raises the importance of the study of social robustness in the face of environmental change. Increasing robustness against a range of environmental changes is a highly attractive strategy because it bypasses the difficult problems of predicting long-term environmental and social change. However, little is known about what makes social, economic, and technological systems robust, and the concept itself needs much more careful conceptualization.

The importance of the problem is suggested comparing two plausible arguments, both found in this chapter. One is that expansion of the market increases robustness by giving economic actors more flexibility in providing for their needs. This argument implies that further penetration of international markets will make it easier for humanity to withstand global changes without major suffering. The other argument is that sociocultural systems often provide a safety net for individuals, for example, through the obligations of others to provide. Sometimes, as in the responses to drought in northern Nigeria, these two arguments seem to support each other: the sociocultural systems there relied on the availability of urban wage labor as a supplement to subsistence agriculture. But sometimes, as with Amazonian deforestation, the two arguments seem to conflict: wealthy economic actors following market incentives crowd out peoples who have developed flexible sociocultural systems, leaving them neither land nor paid labor. Careful comparisons of cases such as the Sahel and the Amazon might begin to clarify the role of markets and of various sociocultural systems in making social groups more or less robust with respect to environmental change.

HUMAN RESPONSE CAN INVOLVE INTERVENTION ANYWHERE IN THE CYCLE OF CAUSATION

Human responses to global change can involve a variety of interventions of quite different types. It is reasonable to suppose that it makes a difference where an intervention occurs, but there is no body of knowledge that clarifies what different effects are likely to arise from different kinds of interventions. Consider an example in terms of Figure 4-1. To respond to the threat of global warming, a government may regulate automobile manufacture or use (affecting a proximate cause—type P mitigation), institute a variety of fossil fuel taxes or incentives (to affect human systems that drive global change—type H mitigation), support research on solar energy (a more distantly type H mitigation), or support adjustment by investing in a fund to compensate citizens after the warming begins to affect what they value. Many arguments can be raised for each strategy. One may argue that mitigation directed at proximate causes is less likely to have disastrous side effects because it is targeted to the desired change only—or one may argue that adjustments are less likely to have disastrous side effects, for the same reason. One may argue that investing in solar energy is wiser than the other mitigation alternatives because it goes to the root of the carbon dioxide problem—or one may argue that it is less wise because too many things must go right for the investment to succeed. At present, not enough is known to shed light on such arguments in any systematic way.

We doubt that a general theory will be developed any time soon that can specify from the class of an intervention its likely effect and the types of unexpected consequences it might have. Such a theory will probably have to be inductive, and the necessary knowledge base does not exist. It is worthwhile to begin collecting the knowledge now.

Research Needs One research priority in the near term should be to support studies that compare interventions at different points in the same causal cycle to identify their main and secondary effects. For example, the effects of regulating automobile fuel economy (a type P mitigation of global warming) can be compared with the effects of taxing gasoline (a type H mitigation); the effects of drought relief payments (an adjustment) can be compared with systems of crop insurance (an intervention to increase robustness). When the relevant interventions have been tried, the studies should be post hoc; when they have not been tried, theoretical analyses or studies based on responses to hypothetical situations will have to suffice.

Even absent a general theory of human intervention in environmental systems, the variety of opportunities to intervene implies an extensive agenda for "normal" social science research to assess the outcomes of interventions in response to anticipated or experienced environmental change. Research approaches developed for evaluating policy outcomes, studying the implementation process, comparing alternative approaches to regulation, and assessing the environmental and social impacts of government programs and policies can all be readily applied to the assessment of potential or actual responses to global change.

HUMAN RESPONSES AFFECT THE DRIVING FORCES OF GLOBAL CHANGE

Because the relationships of human systems and environmental systems are those of mutual causation, all human responses to global change potentially alter both systems. For many interventions, the secondary effects will be minuscule, but it is not always obvious which interventions will have the minuscule effects. Therefore, as a general rule, our conclusions about research on human causes apply equally to research on human responses. For example, policies in response to global change, which often attempt to change technology, social organization, economic structures, or even attitudes, contribute to the interactions of the human driving forces. Like the human causes, human responses can have short-term and long-term effects that may be quite different. And as with the study of the human causes, the study of human responses must be an interdisciplinary effort. Researchers will have to be attracted to the field from their home disciplines, and interdisciplinary research teams will have to be built. Human responses need to be studied separately at different levels of analysis and at different time scales; comparative studies in different social and temporal contexts are necessary; and research is needed to link responses at one level to those at other levels and short-term effects to long-term ones.

NOTES

1 An intermediate case is that in which people make anticipatory responses based on the experiences of others with similar environmental changes.

2 Systems of distinctions regarding human interventions with respect to hazard are, of course, somewhat arbitrary. As noted by Hohenemser et al., (1985), much finer distinctions are possible than those offered here. Our distinctions are offered as a nearly minimal set for studying the human dimensions of global change. They reflect current usage in the global change research community (for instance, researchers tend to use the term *mitigation* to refer to interventions in the human causes of global change but not to interventions in the consequences), and they emphasize the importance of feedbacks between human responses and human causes of global change.

3 Although the policy debate is usually phrased in terms of global warming, the greenhouse gas emissions that are at the center of the debate do much more to climate than raise the earth's average temperature. In fact, some of the other effects, such as on patterns of precipitation, frequency of major storms, cloud cover, and frequency of extreme-temperature events, may be much more significant in terms of human consequences than changes in average temperature. When we refer to global warming, the reader should understand the whole collection of climatic changes associated with the greenhouse effect.

4 A somewhat different distinction between adaptation and adjustment is sometimes found in the literature and is particularly useful for policy analysis. Adjustment is defined as what any affected system does after it feels the effect of an environmental change; adaptations are actions taken deliberately, before the environmental effect is felt, to make adjustment less difficult, costly, disruptive, or painful. This distinction separates a class of policies (adaptation) from the effects of those policies on people or natural systems (their adjustments). Sometimes, the term *adjustment* is restricted to what affected systems would do in the absence of policies of adaptation, as part of assessments of the benefits of those policies. In terms of the distinctions in this report, blocking, improved adjustment, and improvements in robustness may all be the aims of policies on the response side of global change.

5 The literature on responses to natural hazards distinguishes between *adjustments*, short-term activities such as warnings or evacuations, and *adaptations*, long-term social changes that would lower the cost of a recurrence of a hazard. Some of what that literature considers adaptations are treated here as type H mitigations or interventions to increase robustness. Nevertheless, the natural hazard literature includes a more differentiated typology of adjustments than is presented here, including avoiding the loss (e.g., migration), sharing the loss (e.g., relief assistance,

disaster insurance), and bearing the loss (see White, 1974; Burton et al., 1978). Because this literature focuses on hazards over which humans have very limited control, it understandably offers a more detailed typology of adjustments than of mitigations.

6 *Robustness* is one of a number of related concepts, all of which need more careful conceptualization and analysis in relation to global change issues. *Resilience* often refers to the property of returning to a previous state after being altered by changes in the environment. *Resistance* often refers to the property of remaining unchanged in the face of changes in the environment; *vulnerability* often refers to the opposite— the characteristic of being easily affected by perturbations in the environment—whether or not the system returns to its previous state. We use the term *robustness* for its connotation of continued health in the face of environmental change. But *health*, in human systems, can be a subjective term, with some individuals favoring minimum deviations from preexisting states and others desiring permanent changes in particular directions. Thus, whether it is better for human systems to be resistant in the face of environmental change, or vulnerable but resilient, or vulnerable to certain kinds of permanent alteration, is a value-laden question. But it is one that can be informed by analysis of the different ways human systems change when their environments do.

7 There is considerable controversy in the literature, both with respect to biological and human systems, over the question of whether diversity tends to produce stability or vulnerability in systems. The effect of diversity may well depend on the definition of stability used or on other factors, such as the spatial or temporal scale of analysis being made. The examples used here are not meant to exemplify general conclusions about diversity and stability. For more detailed discussions of the issue in the biological context, see Elton (1958), May (1973), Pimm (1982), and Kikkaw (1986). For extensions to social systems, see Holling (1986), Timmerman (1986), and Liverman et al. (1988).

8 It is, of course, possible for humans to replant tree seeds if the climate shifts faster than tree species can naturally migrate. However, because of sensitivities of tree species to soils, photoperiods, and the presence of other species, artificial migrations of this sort may not be sustainable. Survival of tree species depends on a favorable ecosystem, and ecosystems may not migrate well.

9 The reciprocal of this ratio, gross national product per unit of energy demand, is a measure of the economic productivity of energy.

10 This formulation is indebted to Robinson (1989).

11 Our account of this comparison draws heavily on an analysis done by Mortimore (1989) for the committee.

12 This section draws heavily on the much more detailed discussion of the relations of decision theory to global climatic change by Fischhoff and Furby (1983). Additional provocative ideas for research can be found there.

5
Problems of Theory and Method

Knowledge about the human causes and consequences of global change is incomplete, but not only because of gaps in research. Global change issues pose difficult problems of theory and method that have not been adequately addressed, in large part because social scientists have not been called on to work in areas that pose these problems. An effective research program must therefore address these problems from the start, to improve the foundation for future research.

The theoretical and methodological problems arise from the nature of human interactions with the global environment. These interactions, like the environment itself, exhibit interdependencies and unanticipated consequences; nonlinearities between causes and effects; irreversibility; long time lags; and nested relationships between local, regional, and global activities. In addition, the reflexivity of human activity makes knowledge itself a driving force of the system that is the object of that knowledge. Research on the human dimensions therefore must encompass processes at geographical and temporal scales beyond the range of most conventional social science; forge new links among disciplines; involve researchers who would not otherwise address their attention to problems outside their own disciplines; and manage massive data sets. Novel theoretical constructs and research methods will probably also be needed. This chapter identifies some of the major challenges in the areas of interdisciplinary collaboration, development of theory, and selection of appropriate methods.

INTERDISCIPLINARY COLLABORATION

The relationship between humanity and global environmental change is among the most interdisciplinary of intellectual topics (Chen et al., 1983; Kates et al., 1985). Because the driving forces of global change involve the interactions of various human systems with various environmental systems, and because human responses to global change often affect the driving forces, researchers investigating any one system need to treat other systems as intrinsic to their models. It is not satisfactory, for example, for economic models of agricultural production to assume the continuation of average weather conditions, as they normally do, or for models of ozone depletion to assume that international agreements on the phaseout of CFCs will achieve perfect compliance.

Understanding the human dimensions of global change requires creating bridges between disciplines—both between the social and behavioral sciences and the natural sciences and between the disciplines of social and behavioral science. Interdisciplinary work is not only essential but also potentially beneficial to the individual disciplines. It can improve the quality of the assumptions they make and allow each field to consider applying methods developed in other fields.

WHY GLOBAL CHANGE RESEARCH NEEDS SOCIAL SCIENCE

Global change studies typically make assumptions about at least three aspects of human behavior: what people are doing that might affect the environment (and how that behavior may change over time); how people are affected by changes in the environment (and how their sensitivity to such changes may vary over time); and what information people use (or might use or might desire in the future) in making choices about their relationship to the environment. These assumptions may be made explicitly or they may be embedded in models (e.g., projecting constant rates of increase of population, energy consumption, or CFC production). In either case, the quality of understanding is limited by the quality of the social science on which it is built.

If analysts make erroneous assumptions about how people affect the environment, they may err in estimating rates of environmental change and, perhaps more significantly, by underestimating the uncertainty of their analyses. If they make erroneous assumptions about how the environment affects people, they may neglect feedback processes that might be used to mitigate or adapt to global change. If they make erroneous assumptions about how people use information or about what information they will want, they may misdirect their efforts, perhaps producing information no one needs, or producing information in a way that no one can use it (for instance, failing to provide credible estimates of uncertainty in their analyses, without which responsible action is impossible).

Analysts have often made erroneous assumptions of all three kinds. This is evident in the management of technologies with major environmental impacts, so it may also be true of the management of global change. For example, analysts and public officials have often erred in their attempts to anticipate, interpret, and manipulate lay people's responses to nuclear power stations, pesticides, and hazardous waste incinerators, acting without good information about what motivates or terrifies people about the hazards these technologies present (Fischhoff et al., 1981; Fischhoff and Furby, 1983; National Research Council, 1989b). Acting on such misconceptions can imperil major investments and social relations.

The human behaviors in question are the province of social science. Social scientists have some relevant knowledge and the best idea how predictable and malleable the behavior in question may be. Chapters 3 and 4 note many of the areas in which relevant knowledge may be found, as well as the limits of that knowledge. Social scientists can also contribute to the process of analyzing global change by advising natural scientists about the kinds of information about environmental systems that are needed for decision making. For example, the information usually generated by the soil science disciplines is not of the kind needed to analyze the economic effects of soil erosion; soil scientists could produce the information that economists and policy makers need, given input on the nature of that information.

In addition, social scientists have developed methods that may be useful for developing and validating natural science models of global change processes. For example, social scientists have developed mathematical techniques for comparing and combining imperfect indicators of the same underlying variable to produce more reliable indicators and increase understanding of the sources of disagreement (Bollen, 1980, 1989). Such methods can be employed by atmospheric scientists for estimating models built from scattered or imperfectly reliable data, such as on air pollutants or on relationships between industrial processes and emissions.

Thus, social science can help global change research by improving the inputs to models of the global environment, providing techniques for improving scientific analysis, and soliciting appropriate outputs for decision making and policy analysis. Such benefits have remained potential rather than actual for a long time (Chen et al., 1983; Schneider, 1988). The need for better analysis of the global environment can change the situation, given the proper improvements in the institutional base (see Chapter 7). Global change research can provide the necessary conditions for cross-fertilization by bringing natural and social scientists together to work on projects that require mutual understanding of disciplinary languages and constraints, continuing working relationships, and the development of mutual trust.

WHY RESEARCH ON THE HUMAN DIMENSIONS OF GLOBAL CHANGE NEEDS NATURAL SCIENCE

Social scientists may be tempted to think that human dimensions are their province alone. But when they work on problems of global change, social scientists necessarily make assumptions about the natural world, and erroneous assumptions limit the value of their research. They can make ill-informed assumptions about the importance of particular human actions as causes of global change, about the likelihood of particular environmental changes or the likely rate of change, or about the aspects of environmental change that will matter to people.

If social scientists make erroneous assumptions about the relative importance of different human activities as causes of global change or about the likelihood of particular environmental changes, they risk misdirecting their efforts toward understanding human activities with minimal impact on the global environment or human responses to improbable events, both of which ensure that social science findings will be trivial or irrelevant to problems of global change (see Stern and Oskamp, 1987). If social scientists make erroneous assumptions about the aspects of environmental change that will affect people, they may produce misleading results. For instance, survey research may determine that most people think global warming will not affect them, because they annually experience much larger temperature variations than the 5°C average temperature increase projected (Kempton, 1991). But their reactions may be much different if asked more appropriate questions that recognize that the greenhouse effect not only produces warming, but other, more noticeable effects, such as on the frequency of deadly storms and extreme-temperature events, which secondarily affect the frequency of natural disasters and the availability or price of staple foods. Social scientists need to be careful not to credit natural science projections with greater precision than they have, however. For example, projections of rainfall from global climate models at the level of 300 km grids are highly uncertain and should not be taken uncritically in making projections of the productivity of agriculture.

Confrontation with erroneous assumptions or ignored variables offers some great opportunities for theoretical progress. That promise is most likely to be realized if there is direct and continuing interdisciplinary collaboration. Collaboration provides continuing pressure to attend to the variables favored by each relevant discipline and the opportunity to think about them in an informed way. Thus, it is one thing for an economist to be told that climate affects economic productivity. It is quite another for a climatologist to explain current knowledge and uncertainty in climate projections and for a climatologist and an economist to work together to identify particular climatological variables likely to affect the productivity of particular economic activities. Global change research provides an opportunity to foster such interactions and bring benefits to both social and natural sciences (e.g., Land and Schneider, 1987).

Interdisciplinary Communication in the Social Sciences

The global change research agenda, more clearly than many other topics in social science, demands interdisciplinary cooperation. Chapter 3 makes clear, for instance, that the driving forces of global change involve interactions among the favored variables of all the social sciences. The potential for cooperation exists. Environmental problems have already generated important interdisciplinary contributions in small subfields, such as environmental perception, natural hazards studies, environment and behavior, human ecology, and resource management, often focused on policy questions about land management, energy conservation, and management of natural and technological hazards. Global change research may offer an occasion for the broader development of environmental social science, if special efforts are made to involve researchers from several disciplines in continuing collaboration on common projects (see Chapter 7).

Interdisciplinary collaboration on global change issues may also yield increased understanding of how the social and behavioral sciences relate to each other. For example, different social variables may have different explanatory roles with respect to global change: some important for understanding individual and small-group behavior, some for regional or nation-state analysis, and others for understanding the global picture; some important for explaining immediate responses and others for long-term trends; some important for explaining the behavior of human groups, and others for the differences between groups; and so on.

PROBLEMS OF THEORY CONSTRUCTION

One can appreciate the theoretical challenge of the human dimensions research agenda by comparison with the issues raised in modeling the global atmosphere. There is a broad consensus that general circulation models (GCMs) and related physical and biological models are of great use for improving our understanding of global change. This is true partly because there is reasonable agreement on the general form of the models, if not on the details of what should be included in a model and how submodels should be linked. Physical and biological processes are modeled with difference, differential, and accounting balance equations representing stocks and flows of physical entities. Consider for example, the equation for the total energy content per unit mass of moist air (Q):

$$Q = V^2/2 + gz + cT + Lq$$

where V = wind velocity, g = acceleration due to gravity, z = height above sea level, c = specific heat, T = temperature, q = specific humidity, and L = latent heat of condensation. This equation is typical of the types of relations that enter GCMs. Some quantities, such as g, c, and L, are constant. They represent key parameters in physical laws and remain unchanged within a simulation and across all simulations using this equation. Other quantities are clearly variable, such as V, z, T, and q, although in some analyses, for simplification, they might be held constant within a given simulation. Generally speaking, the equations that link quantities in the model are invariant. Within a model the form of the equation does not change across space or time: no variables are added or dropped, and the functional form linking the variables remains unchanged.

These elements of model structure represent an understanding of the physical and biological world. All models are simplifications of reality, and one of the most important means of simplification is to assume some quantities constant and some or all structural relations invariant. Over the time scales of interest in modeling global environmental change, strong assumptions of uniformity apply. While variables vary and to some degree even parameters may change, the set of relevant variables and the laws linking them do not. This is at the heart of the modeling process. Without the assumption of invariance over the scope of the model, models become too complicated to be useful.

Such invariance assumptions are very troubling when modeling the human dimensions of global change. To model phenomena at any given time scale, it is necessary to distinguish rapidly changing variables, which may be assumed to have reached equilibrium within the time scale, slowly changing variables, which may be treated as constant, and a middle range of variables whose values are among the central concerns of the analysis. The difficulty is that for human systems and decades-to-centuries time scales, there is little consensus on which variables fall in which class. Although nothing precludes the application of various social science simulation models to the 50-100 year time frames appropriate for understanding the human dimensions of global change, we are skeptical about the utility of such exercises because the structural relations assumed constant in a model may be those most changed by phenomena being studied. The models will produce results and may be able to accurately simulate systems for short time intervals, but they will not provide useful information or understanding for longer periods during which major social change may occur. The problem increases when it is likely that social change will be stimulated in order to alter processes in the environment.

This situation suggests various theoretical needs that must be met before simulation models or other formal methods are used to project the human activities that generate, or respond to, global environmental change. These theoretical needs are for concepts and analytical tools for understanding human–environment relations at particular levels of analysis and time scales, and for connecting different levels and time scales. We are not calling for grand theory in this area, such as has been attempted by some scholars in the past. If such grand theory is possible, it must be built on much more detailed analysis of particular human–environment relationships than is now available.

The main theoretical needs relate closely to distinctive characteristics shared by human and environmental systems; interdependencies and unanticipated consequences; nonlinearities between causes and effects; the potential for irreversible change; long time lags; and interactions of smaller-scale systems within the global system. The problem of reflexivity in human activity adds a unique theoretical challenge: because people respond to their own understanding, research on global change is itself an influence on the human response.

Social and behavioral relationships relevant to global change need to be specified both at the global level and in contexts localized in space or time, so that comparative study can show the conditions under which local patterns occur and change. The relationships are largely dynamic ones. For example, changes in the demand for a good can lead to changes in its supply, leading in turn to changes in demand. Modeling such processes, even in a relatively self-contained domain, can be very complex, but it is made more difficult in the case of global change by interdependencies with phenomena from other domains. For instance, fossil energy consumption is a complex function of technology, policy, economic activity, social and political structure, and other variables, which are the provinces of a variety of scientific fields, as is evident in our case studies of energy use in China and the United States.

Research Needs

An interdisciplinary approach that seeks to identify relationships among different types of causal variables—such as prices, beliefs, political-economic systems, geographic dispersion of populations, and technological stock—has the potential to illuminate critical interdependencies and to show how changes in one aspect of technology or society may have unanticipated consequences over a period of time. As the case study of China indicates, comparative studies using data from different political systems can do much to specify the relationships governing energy demand. As the case study of the causes of CFC production shows, longitudinal studies within single countries can show how transformations in energy use patterns can follow from changes in technology and settlement patterns that appear to lie outside the energy system. Taken together, comparative and longitudinal studies can illuminate not only local interdependencies but the factors responsible for observing different patterns in different countries or at different times. Because of the interdisciplinary nature of the social phenomena that drive global change, research on the interrelationships among these phenomena has significant potential to improve social science modeling over the long term.

UNDERSTANDING NONLINEARITIES

On the time scale of centuries, major discontinuities in societies are the norm (e.g., North and Thomas, 1973; North, 1981). Very few nations have had the same type of government or the same national boundaries for the past 200 years. Many of these discontinuities can have implications for the global environment. Although this is obviously the case for major wars, discontinuous peaceful change can also have significant effects. Consider the recent example of political changes in Eastern Europe. Evidence presented in Chapter 3 suggests that market economies are much less energy-intensive than state-socialist ones; if this relationship is reliable, the rapid change in Eastern Europe may have significant implications for the global environment.

Research Needs

Although it is hard to imagine progress in social science sufficient to predict the timing of social revolutions, it is possible to imagine a growing ability to predict the probability of such changes. The issue of major social nonlinearities could be approached by working first to identify those major political-economic transitions that have had significant effects on the global environment and then to expand knowledge about the conditions under which such changes occur.

UNDERSTANDING SOCIAL IRREVERSIBILITY

Some changes in human systems seem to be irreversible. Societies that have developed a stable agricultural economy do not revert to a hunting-and-gathering system; industrial economies may decline but do not seem to revert to preindustrial forms; scientific discoveries rarely come undone; new crops or technologies, once proven, do not disappear from the scene. The rise of an automobilized society in the United States, with its infrastructure of roads, homes, and workplaces that depend on roads for access and powerful political interests organized to maintain and extend that infrastructure, may also be essentially irreversible. If the change is irreversible, the implications are profound for energy demand and therefore the global environment, not only in the United States, but also in other countries that aspire to an American style of economy.

Research Needs

It is important to learn about how major, long-term social transformations affect the global environment; the extent to which they can be reversed, slowed, or redirected; and the conditions under which such changes in trajectory are possible.

DEVELOPING APPROPRIATE ANALYSES FOR THE TIME SCALE OF DECADES TO CENTURIES

Analysis of the human dimensions of global change requires a theoretical structure capable of addressing varying time scales, particularly the longer ones that correspond to processes of physical and ecological change (Clark, 1987). However, as a rule, the behavioral and social sciences have focused on phenomena occurring on time scales from milliseconds (e.g., processes of human visual perception) through decades (e.g., adjustments to capital stock, changes in governmental institutions). They have devoted less systematic effort to explaining events on the time scale of decades to centuries, although there are important exceptions in anthropology (e.g., White, 1959; Steward, 1955, 1977), history (e.g., Braudel, 1983, 1984, 1985; Goldstein, 1988; Tilly, 1989, 1990), economics (e.g., Rostow, 1978; Kuznets, 1983; North and Thomas, 1973), geography (e.g., Chisholm, 1982) and political science (e.g., Modelski, 1987). The explanatory variables typically used in the social sciences are not necessarily applicable to the longer time scales. A focus on decades to centuries appears at first glance to favor some explanatory variables over others, and possibly some social science disciplines over others. Several explanatory variables in social science seem immediately applicable to the time scale of decades to centuries:

1. *Demographic shifts* Fertility, mortality (before completion of childbearing), and migration have predictable effects on the sizes of populations over many decades (Lee, 1978; Lindert, 1978). Urbanization and suburbanization take decades to occur and may be stable over much longer periods if they become embodied in long-lived buildings and supporting economic and political structures.

2. *Investment*   Purchases of manufacturing equipment and consumer durables are "built in" for a decade or two; investments in buildings, roads, and water supply systems may last centuries.

3. *Socialization*   Intergenerational learning, by definition, takes decades to change. Environmentally relevant attitudes may be shaped mainly by personal experience or by socialization from the previous generation, with different consequences for the maximum rate of change of such attitudes in a population.

4. *Major global social transformations*   The rise and spread of capitalism, the industrial revolution, the decline of the European peasantry, the development and transformation of the nation-state system, the settlement and development of frontiers, and the creation of a global market all occur on the time scale of centuries (Cronon, 1983; Polanyi, 1944; Wallerstein, 1974, 1980, 1988; Wolf, 1982). Many of them are closely correlated with anthropogenic global change, and some are probably causative.

5. *International regimes*   Patterns of formal and informal practice among nations evolve on the time scale of decades or more, and some are relevant to environmental management (e.g., Wallerstein, 1974, 1980, 1988; Cox, 1987). Examples include the emerging international norm that makes states responsible for environmental protection; the development of international organizations, such as the International Atomic Energy Agency; and changing practice at the World Bank with respect to environment and development.

6. *Family and labor force structure*   Changes in household size, female labor force participation, and educational levels of adult populations occur over several decades (Ryder, 1965; Lindert, 1978). Such changes can indirectly affect the global environment through impacts on economic development, energy demand, and population growth rates.

7. *Diffusion of innovation*   Change in technology for manufacturing, commercial, or consumer use; practices in agriculture, commerce, or government regulation; and the design of social and political institutions normally takes a few decades or more (e.g., Sahal, 1981; Ausubel, 1989). The social time lag between the development of innovations that can affect the global environment and their implementation depends on conditions not yet well understood.

8. *National social transformations*   Even revolutionary changes, such as the ascendancy of market economic principles in Poland and Hungary in 1989 or the Reagan deregulation policies in the United States in 1981, often take decades for their full effects on the environment to appear because of resistance to implementa-

tion. Enforcement mechanisms, regulatory institutions, and bureaucratic procedures all tend to change quite slowly; indirect effects through impact on other countries are even slower.

Research Needs

Global change studies can benefit greatly from focused studies of the sources of variation in those slowly changing aspects of human systems that have major environmental impacts. Other social variables that affect the environment, but that operate on much shorter time scales, also deserve attention. These include individual judgment and choice, social influence, attitude formation and change, and noninvestment expenditures. Under many conditions, it may be reasonable for analyses using longer time scales to treat such rapidly changing variables as if they were constants, assuming that aggregating them dampens variation. However, it is important to understand the conditions under which these rapidly changing factors can vary systematically over time or space, and therefore should not be treated as constants. For instance, the proportion of consumer spending devoted to vacation travel, which intensively uses fossil fuels, can vary between countries, or over time, as a function of overall income level, the proportion of households with small children, or other factors (Schipper et al., 1989). Small changes in this aspect of time use can have a large multiplier effect on long-term projections of greenhouse gas emissions.

ANALYZING THE SPATIAL SCALES OF HUMAN ACTIVITY

To adequately address the human dimensions of global change will require analyses at the global scale and at smaller levels of aggregation. Global-level analysis searches for explanations with worldwide applicability; lower-level analysis presumes that the human causes and consequences vary significantly by region or place. The latter style of analysis is exemplified by Soviet studies of anthropogenic landscape modification, which have examined in detail the regional interactions between human activities and biogeochemical processes and underlying landscape changes (Kotlyakov et al., 1988; Mather and Sdasyuk, 1990).

Scale issues are important because the world community will demand a foundation of knowledge for tackling environmental problems at all spatial scales. Scale issues also raise two important theoretical needs: for theory about human interactions with the environment at the global scale, and for theory to relate smaller-scale activities to the global scale.

Global and continental-scale analyses are needed because social science has done relatively little work at those scales and because so many important human–environment processes have global causes. The theoretical needs can be illustrated with the example of the causes of aggregate global deforestation. Relevant variables surely include population and technological change, levels of and inadequacies in market development, numerous aspects of socio-economic organization, and national policies (Clark, 1988; Turner, 1989). These variables are typically measured at the national level and globally aggregated. But other variables at the truly global level of analysis may also be related to deforestation: global industrialization, market penetration, and flows of investment and information are examples.

Analysis at the global scale may be needed even for local responses to environmental problems. For example, regulation of localized industrial pollution can diminish a country's attractiveness to international investment; the depletion of resources at one locale can increase pressures on other sources of supply; local environmental disasters or degradation may prompt migration to other areas.

Events at lower levels of spatial aggregation are also significant for global change studies. Human–environment studies at the scales of regions or places, focusing on nation-states, firms, social groups, and individuals can show how specific sites and situations affect the ways in which the earth is sustained, altered, or transformed and the ways humans are affected by global change. They illustrate the unevenness of the processes and impacts of change, even systemic change.

The regional approach is important for linking analyses at lower levels to global processes. For example, population pressures and market-based demand, which vary in strength across the globe and by resource and environmental setting, can have global environmental effects. Albedo changes in the North American Great Plains are, in part, a response to agricultural land use changes that are, in turn, influenced by national and international (but not local) agricultural demands. Albedo changes in the West African Sahel, by contrast, are a response to land use changes created by the dynamics of international markets and local subsistence needs, but not national agricultural demand. The pressure of growing populations on local resources can often be traced to the diffusion of medical and public health technologies that have lowered birth rates. Although, in some instances, local or regional relationships will parallel the global, in others, nested sets of explanations will be required that fit local conditions to the regional and the regional to the global. These types of connections must be articulated to understand the human dimensions of global change and to match on the social side the sophistication of physical science inputs to global change research.

Research Needs

Much research is needed to clarify the workings of social factors that operate at the global level and the strength, directness, and spatial scale on which they interact with environmental change. Studies focusing on global-scale variables can illuminate interdependencies and interactions that may not be evident from analyses at smaller spatial scales, even if they are aggregated globally. Also needed are regional comparative studies that identify both similarities and differences between regions in relations between human activity and environmental change, such as the role that population growth plays in deforestation under different socioeconomic, political, and technological conditions. Worldwide comparative studies of this sort have been unusual in the social sciences, and particularly in studies of human–environment relations. International comparative studies are common in environmental policy research, however (e.g., Vogel, 1986; Jasanoff, 1986; Jasper, 1990), and steps in that direction are beginning to be taken in other fields of environmental social science (e.g., Rudel, 1989; Turner and Meyer, 1991). In addition, there is need for studies below the global level designed to assess the possibility that critical human–environment relationships, and the identity of the most important variables affecting those relationships, may vary with the spatial scale of the analysis (Chisholm, 1980, 1982) and to specify the links between levels of analysis.

DEALING WITH THE PROBLEM OF REFLEXIVITY

Reflexivity is important for global change research because understanding global change can alter the global environment itself—not, of course, through any physical properties of knowledge, but indirectly through effects on human activity. The social effects of research on CFCs and the ozone layer illustrate the issue. In this instance, knowledge promoted international cooperation aimed at mitigating global change. One can also imagine knowledge producing conflict, with quite different environmental consequences. If general circulation models improve in their ability to make regional projections of temperature and rainfall, some countries may conclude that continued emissions of greenhouse gases would help their agricultural economies, while other countries would see themselves as losers. This sort of knowledge might stand in the way of international cooperation to reduce greenhouse gas emissions and thus exacerbate the problem the knowledge identified.

Although reflexivity has long been of theoretical interest in social science (Mead, 1934; Merton, 1949; Haas, 1990), problems of global change give it new practical importance. Choices of research agenda, which determine what new knowledge may develop, and decisions about how to communicate that knowledge significantly affect how people respond to global change. Knowledge and experience about these processes can be found in studies of the role of scientific knowledge, expertise, and communication in technological controversies (e.g., Jasanoff, 1990; Mazur, 1981; Nelkin, 1979, 1988; National Research Council, 1989b; Mileti and Fitzpatrick, 1991) and in work on scientific and medical ethics (e.g., Dyson, 1979; Beauchamp and Childress, 1989).

Research Needs

There is much room for additional analysis of past experience with reflexivity of scientific knowledge and for discussions of ethical issues raised by the power of scientific agendas to change human–environment relations.

SELECTING APPROPRIATE METHODS

This section first addresses the issue of appropriate methods for the basic science of human interactions with the global environment—that is, for understanding the nature of the relevant human systems and their interactions with environmental systems. It then turns to the corresponding applied science—the development and proper use of this understanding for informing practical choices among ways to respond to global change.

METHODS FOR IMPROVING UNDERSTANDING

Strategic planning for global change research, both in the United States and internationally, places strong emphasis on integrative modeling of earth systems, including both the environmental and the human (e.g., National Research Council, 1990b). This strategy has been greatly influenced by the example of general circulation models. The GCMs are particularly influential in generating public and policy concern because they can suggest what will happen and can be used to examine alternative scenarios of human action. Many scientists and policy makers hope that models with these attributes can be continuously improved through the use of new submodels, new data, and better computing hardware and software, so that increments of research effort will yield improved global projections.

Emphasis on integrative modeling is reflected in general systems diagrams such as that displayed in Figure 5-1. These diagrams have great utility in ordering the elements of global change, identifying links among processes, and suggesting critical connections among research agendas. The diagrams usually identify subsystems for human activity. For example, Figure 5-1 includes human activities as both a driving force that produces three outputs (land use, CO_2, and pollutants) and a causal link between climate change and changes in the proximate causes of global change. One can construct a plausible research agenda by thinking through the knowledge that would be required to build mathematical models of each of the subsystems represented by boxes in the figure (including an expanded number of boxes to replace the single one labeled *human activities*) and to link the models to each other. In fact, some commentators on the research agenda for global environmental change seem to give an important place to the task of building and linking models such as those implied in the diagrams (National Research Council, 1990b).

While the systems diagrams provide useful heuristics, we wish to sound a warning against overstressing formal modeling, particularly at this stage of research on human dimensions. Understanding the dynamics of the general processes of land use and those that produce CO_2 and pollutants does not translate readily into mathematical models at current levels of knowledge. Serious unsolved conceptual problems are implied in our discussions of the problems of theory construction: the shape of the relationships is subject to change; such change may even be a goal of policy; and much more needs to be known about interdependencies across domains, issues of nonlinearity, irreversibility, and reflexivity, and the relationships between events at different levels of temporal and spatial aggregation. Because work on these

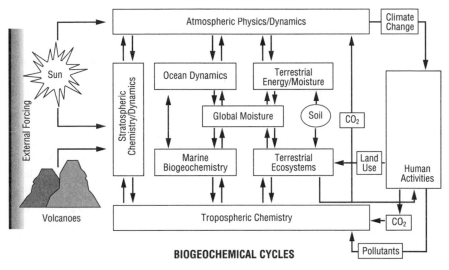

FIGURE 5.1 Status of earth system science in the year 2000. Source: National Research Council, 1990b.

issues is so little developed, global models of human–environment interactions can provide only highly uncertain projections of the future.

There are, however, useful analytic alternatives to grand mathematical models. It is impossible to simultaneously maximize generality, realism, and precision in a model (Levins, 1966; Puccia and Levins, 1985). Levins suggests three options: (1) sacrifice generality to realism and precision; (2) sacrifice realism to generality and precision; or (3) sacrifice precision to realism and generality.

The first approach of sacrificing generality for realism and precision is typical of most simulation models developed for policy analysis. These models usually include dozens, hundreds, or even thousands of equations describing linkages among variables. They are realistic in the sense that they reflect the detailed structure of the social system being modeled—different equations describe each sector of the economy, each segment of the population, each region in the country or the world. Precision lies in the specification of relations among variables through the equations. But the detail that provides realism forces a loss of generality—the models must be developed for each specific application, and work with a new region or problem may require extensive recalibration of the model. For analyses with short time horizons, such models may achieve their goals of realism and precision. For example, they might be appropriate for understanding the short-term impacts on a local or national economy of a policy to reduce CO_2 emissions. But we believe they are not realistic nor likely to be precise for longer time horizons because the social, political, and economic relationships represented by the equations in such models are likely to change over the longer term. Thus we do not see the extensive use of such models as an optimal strategy for analyzing human response. During the last decade, there has been a rapid growth in scholarship that critically assesses the technical features of such models, and this may lead to better formulations that have a more dynamic form, or at least to a wider appreciation of the limits of these models (Belsley and Kuh, 1986; Brewer, 1983; Greenberger, 1983; Greenberger et al., 1976; Kmenta and Ramsey, 1982; Meadows and Robinson, 1985; Stern, 1984, 1986).

The second approach, sacrificing realism to generality and precision, is typical of simple, heuristic, analytical models. These models usually consist of one or a few equations that are intended to capture key elements of the behavior of an admittedly very simplified system. They are analyzed using standard mathematical tools or, if intractable, are simulated. They are not realistic in that they greatly simplify the system being analyzed. Like models of the first type, their mathematical formulation gives them precision. Their relative simplicity enables them to be used in a variety of circumstances, precisely because they do not incorporate the details that typify models of the first type. Of course, these first two types of models are not distinct, as simulation models are usually built with a large number of simple analytic models linked together. Historically, the analytical models have had the disadvantage of making simplifying assumptions that presumed equilibrium or stasis, and so are not ideal for studying long-term responses. Recent work with developing nonequilibrium and evolutionary models suggests ways past these assumptions by offering model structures that are much more dynamic in character and that allow for changes in both parameters and model structure within the model (Axelrod, 1984; Boyd and Richerson, 1985; Holling, 1986; Smith and DeJong, 1981). We believe that these new methods for analytical modeling may prove very fruitful in addressing the kinds of structural transformations that are likely to accompany global change.

The third approach, which sacrifices precision for generality and realism, is typical of a good deal of traditional social theory in which the models are informal, exploratory, and—a key feature of the type—nonmathematical. These types of models have many faults: they often do not produce clear forecasts for assessing policy options; they may be imprecisely specified; they do not readily link to other models from the social, physical, and biological sciences; and, because of their imprecision, the full implications of a specification may be unclear. Still, much social science knowledge is based on such models, and we believe they will continue to be fruitful despite their flaws. Because they are flexible and closely linked to the kinds of data available, they can help inform the other two approaches. We suggest that efforts to understand the human dimensions of global change can make fruitful use of such informal models, and that the link between analytic models and qualitative models can be strengthened with analytical techniques that more closely match the character of most qualitative models (Puccia and Levins, 1985). Qualitative modeling in turn can inform the development of newer approaches to analytic modeling. And while we are cautious about the use of simulation models, further methodological work on such models, coupled with the incorporation into them of more dynamic analytical approaches could significantly improve understanding.

Over the near term, research on the human dimensions of global change will best proceed along several parallel tracks, using many methods. Along with quantitative modeling approaches, it will be critical to proceed with studies using other traditional social science methods and aimed at more accurate specification of the relationships that models represent quantitatively and of the conditions under which those relationships develop and change. We conclude, therefore, that at least for the near term, the strength of emphasis on building integrative models that marks other parts of the global change research strategy is premature for studying the human dimensions. For the human interactions research agenda, much more understanding of the underlying processes needs to be developed before great strides can be made in integrative modeling.

The catalogue of methods used for learning about social processes, and the strengths and weaknesses of each, are well known to social scientists. They include randomized experimentation, which can convincingly demonstrate causal relationships, but does not deal well with complex relationships among variables not subject to manipulation; quasi-experimentation, which is useful for assessing interventions in the environment when random assignment to treatments cannot be accomplished; case study, which may use historical, ethnographic, or other methods and which allows consideration of complex relationships but leaves open questions about generality; directed case comparisons, which address the issue of generality by comparing carefully selected situations; multivariate analysis of existing data, which can quantify complex relationships but lacks the richness of the comparative case method; and survey research, which can collect new data on multiple variables in a standardized way and generalize by sampling.

One can make reasonable judgments about the appropriateness of each method for studying particular questions. Different methods are more or less appropriate depending on the level of spatial or temporal aggregation of the questions asked, the availability of standardized data, the number of variables of interest, and so forth. But for the broad project of global change research, we emphasize the importance of a multimethod approach. Different methods tend to illuminate different aspects of a process, and each method can be used as a check on the results obtained by other methods.

To understand interdependencies in humans and the environment, it is best to promote methodological interdependence. The point can be illustrated by the historical case study of the causes of CFC use in Chapter 3. The case study identified causal connections not often noticed between the development of CFCs and fossil energy demand, operating over half a century. These are represented schematically in Figure 5-2. These connections would have not been revealed by a more standard analysis based only on the study of data on CFC use, regardless of the method used. But once revealed, the connections raise new questions best addressed by other methods. For example, the CFC case underlines the value of more detailed quantitative analysis of relationships between the spatial dispersion of an affluent population and demand for CFCs and fossil energy. This sort of dialogue of methods is likely to lead to a more complete picture of the human causes of global change than a collection of unrelated discipline-based or method-centered studies. The most likely way to bring the results of different methods into contact is probably through research communities united by a common set of problems. Therefore, we emphasize the need to build an interdisciplinary, multimethod research community for the study of the human dimensions of global change. We discuss elements of the creation of such a community in Chapter 7.

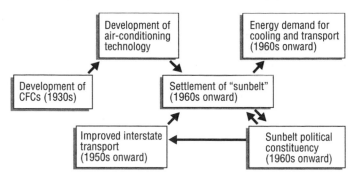

FIGURE 5.2 Social consequences of development of CFC technology for refrigeration, 1931 to present.

METHODS FOR INFORMING CHOICE

The model presented in Figure 4-1 emphasizes that human systems respond to global environmental changes as a function of the way those changes affect things people value. Accordingly, the applied social science of global change focuses on assessing those effects and evaluating each available response option in terms of their direct and indirect effects on what people value. It is obviously difficult to make the necessary assessments and evaluations. Nevertheless, social scientists have developed techniques aimed at assessing the social effects of environmental change and of policy interventions and at placing values on effects of very different kinds, which are not readily measurable on a common scale.

Social Impact Assessment Methodology

A set of methods, known collectively as social impact assessment (SIA) methodology, has been developed that could be adapted to the problem of assessing the effects of global change and the policies enacted to respond to it (Finsterbusch et al., 1983; Dietz, 1987). SIA developed out of the impact assessment provisions of the National Environmental Policy Act (NEPA); it uses some of the methods typical of technology assessment and risk assessment, but it is more focused on the effects on human systems than those activities sometimes are. The experience with SIA illuminates several important issues involved in formally assessing the consequences of global change.

The goal of SIA is to predict the social effects of policies, programs, and especially projects that cause change in the physical and biotic environment and to use these predictions as an aid to decision making and public debate. Thus, SIAs have strong parallels to the kinds of analysis necessary for assessing the proximal effects of global environmental change on things humans value.

Formal Modeling Approaches   A dominant approach in the early days of SIA was to develop and apply computer simulation models for impact assessment (Dietz and Dunning, 1983; Leistritz and Murdock, 1981). The typical model began with predictions of the labor required for the construction and operation of a project such as a prison or nuclear power plant. The model then estimated the demand for secondary jobs, the extent to which the demand could be met locally, the subsequent in-migration, and finally the demand for government services and stresses on the local governmental budget.

These models have several advantages for SIAs. They are based on seemingly hard and objective numbers—such as the relationship between primary and secondary jobs—that in principle can be evaluated theoretically and can be tested with empirical evidence.   They produce quantitative forecasts of the types of impacts that were expected to be of great importance. And they can be applied quickly and easily to alternative versions of the same proposed project or to other projects with the input of a modest amount of situation-specific information. Because of these characteristics, a number of these models saw widespread use, and some are still applied today.

Unfortunately, these models have several important flaws that have led to their declining importance in SIA practice. First, the ability of the models to make accurate predictions is very limited. Despite the use of such models for 20 years, there have been very few attempts at model validation by comparing projections with actual experience.   The evidence suggests that key parameters that are assumed in the models to be constant over time, such as the ratio of secondary to primary jobs, are in fact highly variable over time and space and are quite likely to change drastically under the influence of large scale projects such as those simulated (Meidinger and Schnaiberg, 1980). Thus, these models often have at their core postulated relationships that do not match social reality. For the practical purposes of these models, the lack of predictive validity is a serious flaw.

A second fault of the models is what they leave out. All models simplify the world. These models handled only the impacts that could be represented with fairly simple quantification methods, such as input-output analysis or standard population projection techniques. As it turned out, among the most controversial impacts of large resource development projects was the prospect that they would disrupt community life, resulting in mental health problems, adverse effects on youth and the elderly, and so on. Although concern with such issues produced enough opposition in some cases to slow or block entirely the construction of large projects, these impacts were ignored in the simulation models. The analytical resources used for simulation had been focused on the wrong targets in terms of the political choices at hand.

Finally, the projections of such models often acquire the status of facts. Modelers are usually sensitive to the conditional character of projections and the need to use them heuristically. But in practice, a few runs of the model can become fixed in the minds of those debating a proposed project, program, or policy, with other important issues pushed aside. People who do not find the models addressing the issues of most concern to them become skeptical of the entire analytic approach, with the result that scientific analysis becomes a lightning rod for debate rather than a way to clarify the issues.

Formal models applied to the consequences of global environmental change may have analogous uses and problems. The demand for social impact assessments of the anthropogenic effects of global warming is already large and will certainly grow. In response to this demand, we expect to see many well-intentioned efforts to use simulation models. Such models usually take the output of physical and biological process models that are relatively well developed compared with social process models and use them to drive simple economic and demographic models. These simulations have great appeal because they are easy to develop from existing models of component systems and because they produce seemingly hard results. However, they have at least two serious limitations, in addition to the inherent difficulty of assessing the sorts of consequences SIAs often leave out.

One problem is that SIA methods were developed to assess local impacts of environmental changes and human interventions, but much of the need is for global assessments. SIA methods, taking into account the limitations already mentioned, are appropriate for assessing the local social consequences of, for instance, a change in rainfall or an energy conservation incentive policy. But they have not been built to assess the *global* impacts of the same events, an assessment that would be necessary for evaluating policies for mitigating global climate change. To do this would require a SIA for each locality affected and a technique for aggregating the SIAs.

The other problem is that SIAs, which take considerable effort and expense, especially if used to assess global effects, are likely to distract attention from more important analytical and modeling questions, including fundamental questions about interactions between environmental systems and human responses. A lesson of the last two decades of SIA that might also become a lesson of applying SIA methods to global change is that simulation models of the sort described here have limited utility in advancing scientific knowledge or in aiding serious discourse on policy: their value may be illusory. This is not to decry all use of simulation, but to sound a warning about the potential for misuse of applied simulations for policy purposes.

Post Hoc Evaluation Approaches The SIA literature includes empirical studies of the impacts of specific projects, programs, or policies after implementation. These studies employ conventional and well-developed social science methods and theory to determine what actually happened in response to public or private action (Finsterbusch et al., 1983). The largest body of this research addresses the effects of energy development boom towns on the community and its members (Freudenburg, 1984). Such studies are very important to SIA for several reasons. First, they pointed out the importance of context in determining impacts and provided a useful caution against overgeneralizing or assuming that a specific local or regional future could be forecast with much accuracy. Second, they encouraged the normal working of empirical science by stimulating debate about the theory, methods, and interpretation of past studies and allowed a variety of methods and theories to be applied to the same general problem. For example, the effects of boom town developments on youth have been studied using ethnographic methods, surveys, and statistical analysis of secondary data. New studies benefited from earlier studies, and theory, methods, and results improved. Finally, well-conceived post hoc studies borrowed from and contributed to theory and method both in evaluation research and in the traditional disciplines, often forcing the disciplines to consider problems they had not previously addressed. In many cases, the researchers carrying out these empirical efforts worked as part of interdisciplinary teams and learned to develop more integrated pictures of social reality than were available in the parent disciplines.

Nevertheless, the achievements of 20 years of empirical work on social impacts have been modest. The work has often been conducted with very limited resources and tight deadlines, with resulting compromises in conceptualization and research design. Long-term, post hoc studies, which would be invaluable for improving prediction tools, are especially uncommon. Millions of dollars have been spent on predicting impacts, but comparatively little on measuring the actual impacts. The situation is rather like trying to understand population dynamics and develop sound population projection techniques in the absence of census data to modify the understanding and calibrate the techniques.

The problem of limited resources is exacerbated by the marginal linkage of impact studies to the prestigious centers of core disciplines. The main streams of most social science disciplines have ignored the effects on society of the physical and biotic environment and of technology. As a result, it has been difficult to obtain support from conventional funding agencies or publication in the most prestigious journals. Many of the post hoc SIA studies have been doctoral dissertations or relatively small efforts funded from intrauniversity sources. Publication has often been in interdisciplinary journals that lack the circulation, prestige, and visibility of the disciplinary journals. The lack of funds and of prestige, in turn makes it difficult to attract the best young talent to these studies.

Research needs: The experience with post hoc SIA studies offers several useful lessons regarding research on human dimensions of global change. Perhaps the simplest but most important is the value of ongoing studies to monitor the impacts of projects, programs, and policies. Although social program evaluation is a sophisticated and well-developed activity, the best methods of evaluation research have rarely been applied to environmental or resource policies or programs. A major exception to this rule is in the area of residential and commercial energy conservation. Post hoc evaluations should be viewed as an important part of the process of analyzing policy alternatives for response to global change, and resources should be provided for them. Even when the expected impacts are too far in the future to be studied directly, empirical methods should be applied to analogous programs or policies.

A second lesson is that, although the standard tools of disciplinary research are of central importance in applied research on the human dimensions of global change, the disciplines will be slow to accept the importance of that research. As a result, the growth of research capability requires lines of support—financial and intellectual—in contact with but somewhat autonomous from the disciplines. We return to this issue in Chapter 7. For the moment, we note that agencies and foundations that support basic research should play a key role in fostering rigorous, theoretically informed, and methodologically well-designed empirical studies on human consequences of and responses to global change. In addition, mission agencies that usually fund only applied social science research, but that have some basic research programs in the natural sciences, should initiate basic research in the social sciences as well.

Methods for Valuation

To have practical importance, any assessment of the human consequences of global change, or of responses to it, must be combined with some means of placing values on the consequences. Valuation is a difficult task because consequences can be of very different kinds, so that a metric for making trade-offs is not obvious. Moreover, different people often place different values on the same consequence.

In the last 5 to 10 years, the problem of valuation, or commensuration of the various things humans value, has become an important theme in the SIA literature (Mitchell and Carson, 1988; Dietz, 1987). The typical impact assessment describes a diverse list of probable impacts, for example, increased jobs, lower energy costs, loss of wildlife habitat, increased cultural diversity in the community, and increased crime and congestion. Often, the impact assessment makes no attempt to assign relative value to these impacts, on the assumption that relative values are and should be assigned in the political process, not by the researchers. In this model of decision making, estimates of impacts are used as inputs to the political process, to better inform the inescapable debates between conflicting values and interests.

Although this approach has dominated most impact assessment, other forms of policy analysis in the social sciences have tried to systematize valuation. The premier technique is benefit-cost analysis (BCA). In BCA, impacts are assigned market values, either directly or by imputation. Future impacts are discounted. Then, having assigned a value to all costs and benefits in current dollars, the ratio of benefits to costs, net present value, internal rate of return, and other measures of the efficiency of the project can be calculated. While those who produce BCAs typically caution against taking the final ratio or other measures of efficiency too seriously as decision criteria, the process does provide an explicit framework for valuing otherwise incommensurable impacts. It also suggests the appropriate action to be taken from among the options analyzed. Some versions of risk analysis, such as risk-risk and risk-benefit analysis, provide similar structures for making trade-offs among impacts (Dietz et al., 1991).

There are many strong criticisms of BCA and risk analysis as decision-making tools for environmental policy (e.g., Dietz, 1988; Fischhoff, 1979; Mazur, 1981). These include methodological critiques and deeper political and ethical questions about whether it is appropriate for technical analysts to decide the relative values of outcomes instead of leaving such trade-offs to the political system. But whatever the difficulties of particular methods and their implementation, they have the advantage of forcing careful and systematic thinking about the appropriate methods for assigning values to impacts and thus making trade-offs among impacts. Recent debates in the SIA literature suggest that much more work needs to be done on the problem of valuation. Several approaches that might serve as complements to BCA, using very different theoretical underpinnings, have been offered (Dietz, 1987; Freeman and Frey, 1986). Some of the alternatives rely on assessing people's value preferences by directly eliciting them (e.g., Keeney and Raiffa, 1976). None of the proposed approaches is yet widely accepted, but a dialogue has begun on the subject. At the heart of the dialogue is a belief that, while social science methods can never replace the political process in assigning values to impacts and making decisions about the allocations of resources, it is equally naive to believe that social science research cannot inform the valuation process. We believe the history of BCA has shown the value and influence of systematic thinking on value issues. Recent work on valuation in SIA is seeking to expand the domain of that thinking.

This suggests a simple lesson for research on human dimensions of global change. Although the social sciences cannot resolve the value questions at the heart of individual and collective decision-making about global change, social scientists should offer their best systematic insights into methods for understanding values, value conflicts, and the implications of alternative approaches to individual and collective choice. Here the logical interdisciplinary collaborations are not only with physical and biological scientists, but also with scholars in the humanities, especially philosophy and the arts.

The Appropriate Uses of Research

Because of reflexivity, research on global change has the potential to seriously affect human societies. It is therefore important for researchers to be clear about the purposes of their work and to develop a context for it that is not politically naive. The problems arise most clearly in the use of formal models in policy analysis, because the connection to social choices is quite direct, but they exist as well with other methods of policy analysis.

The last decade has seen increasing attention to the use of models in policy analysis and collective decision making. Most of that literature has been sharply critical of the effect of simulation models on the policy process (Baumgartner and Midttun, 1987; Brewer, 1983; Freedman et al., 1983; Greenberger, 1983; Greenberger et al., 1976; Habermas, 1970; Hoos, 1972; Meadows and Robinson, 1985; Robinson, 1982, 1988; Stern, 1984, 1986; Wynne, 1984). The problem can be explicated by drawing the distinction between forecasting and projecting. Most modelers would argue that models project rather than predict. If the systems of structural relations incorporated into a model are a reasonable approximation of reality at present and if those structural relations do not change, the model implies that certain events *would* occur. However, as noted above, the structural relations among the elements of social models are quite likely to change in the intermediate and long terms. Indeed, the goal of policy modeling is to assess the effects of intentional changes in the social world. Since the models by necessity simplify the world and assume to be constant that which is not, their predictions can have only limited validity. The exact simplifications and assumptions made can never be fully validated by scientific knowledge, but always depend to a large degree on the judgment of the modeler, and that judgment reflects a modeler's world view as well as objective fact. While the research community understands these limits of all models, in the policy process model results often are abused. The abuse takes two related forms.

First, the results of modeling are sometimes interpreted as what *will* happen—a forecast or prediction rather than a simple projection of the current situation (Ascher, 1978; Brewer, 1983; Baumgartner and Midttun, 1987). Present actions are influenced by a prediction of the future, but the assumptions that influence that prediction are hidden in the details of a model and not easily accessible to debate. This limits consideration of policy responses to those options that fit within the structure of the model and precludes easy consideration of policies that might make structural alterations (Meadows and Robinson, 1985; Robinson, 1988). In this way, the structure of a model can become a determinant of policy, rather than simply a mode of evaluating options. Technical analysis thus drives the policy process (Habermas, 1970; Dietz, 1987).

Second, the results of a modeling exercise have the aura of scientific objectivity, despite the judgments built into them. A model and its results are sometimes erroneously seen as neutral, and the selection of policy options is consequently justified as a technical activity rather than a political one. This perception makes models a powerful tool of advocacy (Majone, 1989). Decision-making institutions claim to be responding to the forecasts in the only logical way, but in fact, "forecasts do not reveal the future but justify the subsequent creation of the future" (Robinson, 1988:338). As noted in the discussion of social impact assessment, this way of using models can either limit creativity and debate or provoke skepticism. In the latter case, the models are ignored rather than being used in an informative role. The political process drives or ignores the technical analysis.

Recently, a number of critics have suggested alternative criteria for modeling that preserve the benefits of modeling without the political problems (Dietz, 1987; Steenbergen, 1983; Masini, 1984; Robinson, 1988). They argue that modeling should be used to increase the range of scenarios considered in the policy process, so that each scenario can be evaluated in terms of its desirability. This is essentially the approach of normative forecasting—the model is used to work backward from a desirable future into present policy options that could create that future. A key element in the approach is the recognition that all models build in assumptions and simplifications. To the greatest extent possible these assumptions and simplifications should be explicit to those using the model *and* those using information from the model. This approach places the modeling exercise within the policy process and explicitly allows interested parties to argue for their own assumptions about reality and for the inclusion of variables of critical concern to them.

We are not opposed to the extensive use of modeling in the study of human dimensions of global change. What is needed is thoughtful modeling, rather than stock application of existing methods, however technically sophisticated those methods may be. Also needed is more clarity about the purpose for using particular models in particular analyses. The challenge of understanding global change provides a valuable opportunity to advance the state of the art in social science modeling. Social models will be most helpful, and best able to link to emerging models in the biological and physical sciences, if they take serious account of the dynamic core of human action and social structure and if they are sensitive to the ways in which models are used in the policy process. Thus we advocate research on the methodology and policy use of models, rather than simply the elaboration of existing models and modeling methods.

Research on global change is likely to become the object of policy discussions regardless of the method used. Formal models are particularly potent influences in some circles of the policy community, but case studies, surveys, econometric analyses, and experiments can also be oversold as revealing absolute truths. Descriptions of the results of natural science can also be politically powerful, because most decision makers get their knowledge secondhand and in predigested form, and because the digested version necessarily embodies judgments in addition to those involved in the original research that the recipient cannot recognize as such (National Research Council, 1989b; on related issues, see Fischhoff, 1989:238-257; Jasanoff, 1990). The challenge for research is to generate knowledge responsive to a range of policy concerns; the challenge for policy is to maintain institutions that bring out the range of responsible interpretations of that knowledge so that individual and collective decision makers can have an informed basis for action (National Research Council, 1989b; Stern, 1991).

CONCLUSIONS

We have noted in this chapter the significant problems involved in building theory and choosing methods that will help improve understanding of the human dimensions of global change and inform human responses. We reached four broad conclusions.

INTERDISCIPLINARY COLLABORATION IS ESSENTIAL

The nature of global environmental change is such that the variables of central concern to a variety of disciplines typically act in conjunction. Consequently, for many important global change processes, specialists in any one discipline are likely to make erroneous analyses if they fail to draw on expertise from other relevant disciplines. In some instances, the necessary cross-fertilization can be accomplished through the usual processes of scholarly debate in academic journals and meetings, but in many instances, this is unlikely to occur because the relevant scholars do not interact. Consequently, the most promising way to get the necessary interaction is within interdisciplinary research groups and communities brought together by a problem of common interest. It should be a high priority of the human interactions research effort to support problem-centered interaction among social and natural scientists, for example, through research projects that require such contact, problem-focused scholarly meetings, and interdisciplinary research centers.

New Theoretical Tools Are Required

Because global change studies are inherently interdisciplinary, and especially because the object of study requires analysis at spatial and temporal expanses much greater than most social scientific theory encompasses, these studies challenge social science to develop new theoretical tools. Among the important questions about global change for which improved social theory is required are these:

—Under what conditions do major national and international changes in political-economic structure occur? At what speed are they propagated?

—Which social changes maintain themselves over long periods of time, which are ephemeral, and which have a long-term increasing impact? How and under what conditions can these changes, especially the expanding ones, be reversed, slowed, or redirected?

—What are the major sources of variation and change in the slowly changing aspects of human systems, such as fertility, socialization, building patterns, family and labor force structure, national policy systems, and international regimes?

—Under what conditions do short-term factors in human behavior, such as individual judgment, social influence, and consumption expenditures, vary systematically over time or space?

—How do human activities that occur on the global scale, such as scientific progress and the spread of markets, interact with the global environment?

—What are the relationships between human activities at the global scale and human activities that occur at smaller spatial scales?

—How are nature–society relationships occurring at one level of analysis (say, local deforestation) related to the same relationships at higher levels (e.g., the nation-state)? What determines these connections?

The global change research agenda requires answers to these questions for only a subset of human activities—those with significant implications for the global environment. But many of these questions, broadly stated, are important for basic social science. This convergence of interests between basic social science and global change research has a positive and a negative side. The negative is that theory needed to understand global change does not yet exist. The positive is that studies of the human dimension of global change have the potential to attract scholars who see opportunities to make important theoretical advances for their home disciplines.

METHODOLOGICAL PLURALISM IS THE MOST APPROPRIATE STRATEGY

At least for the near term, a strong emphasis on building integrative models is premature for studying the human dimensions. For the human interactions research agenda, much more understanding of the underlying processes needs to be developed before great strides can be made in integrative modeling. Despite the attractions of integrative modeling, the committee has concluded that, for the present, support for human interactions research should be concentrated on process studies, with only modest levels of support for integrative model building, until improved data and process knowledge provides a better basis for constructing formal models of the human dimensions of global change. Formal modeling should be treated as one among many methods deserving attention for advancing knowledge of the human dimensions of global change. Models should be part of a dialogue of methods, with several complementary methods being used to give a more complete picture of human interactions with the global environment than any single method can produce.

POST HOC ANALYSES ARE ESSENTIAL FOR EVALUATING HUMAN RESPONSES

The public sector will often be involved in efforts to redirect the social driving forces of global environmental change or to respond effectively to the impacts of global change as they become apparent. Many governments are already beginning to take actions to mitigate ozone depletion and other forms of global change by limiting CFC production, promoting energy conservation, and other means. As with other major public policy initiatives, it is important to evaluate both the intended and the unintended consequences of the policies. What works and what does not work in efforts to control the social driving forces underlying global environmental change? What can we learn from the experience with ozone depletion about effective responses to other types of global change? Can public policy play a major role in reducing the energy intensity of advanced, industrial economies? Are some policy instruments (for example, regulations, charges, transferable permits) more effective than others in curbing the anthropogenic sources of global change?

It has been customary in U.S. social policy to evaluate the effects of policy initiatives and to plan for the evaluation in the budgetary process. But post hoc studies of interventions in the environment have not received nearly sufficient government resources or sufficient involvement from the academic centers of social science. To plan for policy, post hoc studies of human responses to past environmental changes are also essential. Agencies and foundations that support basic research on global change should play a key role in fostering rigorous, theoretically informed, and methodologically well-designed empirical studies assessing the effects of human responses to global change. In particular, **we recommend that all federal agencies that sponsor programs anticipated to affect processes of global environmental change should routinely include in their budgets funds to evaluate the effects of those programs after they have been enacted**.

This recommendation is not a proposal for a new form of environmental impact statements. It is a proposal for data gathering *after* a policy is in place, because for adjusting policies and for long-term gains in understanding, it is critically important to assess actual impacts after the fact. Over time, post hoc studies can lead to the accumulation of a sizable body of empirically grounded propositions.

6
Data Needs

The previous chapters suggest that to build knowledge of the human causes and consequences of global environmental change, advances in theory, method, data, and research infrastructure are all imperative. This chapter focuses on the needs for data. We begin with three illustrative research problems and then proceed to a systematic account of issues of data availability, data quality, and collection needs.

THREE HYPOTHETICAL RESEARCH PROJECTS

CAUSES OF GLOBAL WARMING: A CROSS-NATIONAL STUDY

Global change researchers want to understand the relative contributions to global warming of population growth, economic growth, technological change, and various aspects of national policy and social and economic structure. One way to shed light on the issue would be through analysis of time-series data on each country in the world over an extended period; data series are available for many variables in many countries. Such a research project ideally would use some or all of the following data:

Dependent variables
  Atmospheric gas concentrations CO_2, methane, CFCs, etc.
  Greenhouse gas releases Same gases

Independent variables

National population Total, urban-rural split, agricultural-nonagricultural, birth rates, age distribution

Economic activity Gross national product (GNP); production of manufacturing, mining, agricultural, service sectors; income distribution

Energy demand Consumption of coal, oil, gas, nuclear electricity, hydroelectricity, biomass; demand for transportation, space heating, manufacturing process heat, and other major end uses

Technology Transportation subdivided by technology used (e.g., number of passenger miles by automobile, number of ton-miles by train); capacity and fuel efficiency of industrial boilers, steel furnace, etc.

Prices Major energy sources, labor, agricultural land; interest rate for capital investment

Land use Hectares in wetland crops, dryland crops, pasture, forest; number of cattle; dispersion of population around urban centers

Institutions Distribution of land across agricultural users and by land-tenure system; degree of market vs. nonmarket control over markets for energy, land; environmental regulatory style

These data, even when they are desired only on an annual basis and at the national level, are unevenly available. Some, such as national population, are fairly accurate and available, but too infrequent; some, such as GNP, are available and sufficiently frequent, but of highly variable and sometimes unknown accuracy; some, such as energy use subdivided by economic sector, are sufficiently accurate and frequent for some countries (typically, OECD countries), but not available at all for many other countries; and some, such as the degree of market control of the economy, are unavailable because a reliable index does not yet exist. For most of the international data sets that are available, information on data quality is scanty and of uncertain reliability.

HUMAN CONSEQUENCES OF DEPLETION OF STRATOSPHERIC OZONE

Global change researchers want to be able to monitor the possible effects on things humans value of ongoing changes in the level of stratospheric ozone and to attribute the effects to ozone depletion rather than other possible causes. One way to shed light on the issue would be continuing analysis of a time series of data on values that might be affected by increased UV-B radiation at the earth's surface. These include the productivity of crop and forest flora, human incidence of skin cancers and resistance to infectious disease, and the economic consequences of such changes. Such a research project ideally would use data such as the following:

Health   Data at national level or in selected localities on deaths from skin cancer, incidence of skin cancer, and incidence of selected infectious diseases
Agriculture   Productivity data on selected crops; vegetation density in forests
Economic   Value of selected agricultural products in countries or selected localities; health care expenditures
Stratospheric ozone concentration   Measured values from satellite observation for selected regions
UV-B incident radiation   Measurements at ground level
Policy responses   Enactment of policies to limit CFC releases; implementation of such policies; introduction of education campaigns for avoidance of direct exposure to sun; measures of resultant behavior change

Data on these variables are also uneven. A study of the effects of ozone depletion presents some of the same data problems illustrated by the study on the causes of greenhouse gas emissions, as well as others. For instance, data available summarizing political units (e.g., local or country-wide health data) must be compared with data available summarizing physical areas of map grids (e.g., satellite observations of ozone levels or color of vegetation) and with data collected at selected time points (e.g., incidents of UV-B radiation, surveys of individual beliefs or behaviors). Some data, such as satellite images, are available so frequently that techniques of time sampling must be used. Some data, such as that on human health effects or agricultural productivity, may not be available at the desired geographical level or on the precise effect of interest. Data from diverse methodologies, such as remote sensing, survey research, and health records, are generally not available from the same archives or in mutually intelligible form. And finally, some of the effects of interest are as yet unknown, so the monitoring process must leave room for adding other variables and reanalyzing the data.

ECONOMIC AND NONECONOMIC FORCES CAUSING LOSS OF BIOLOGICAL DIVERSITY

The activities most implicated in the loss of biological diversity are influenced by human demographics, technologies, political systems, economics, traditions, and belief systems. In particular, understanding how individuals, tribes, ethnic groups, and nations regard species and ecosystems and reconciling economic and ecological methods of accounting are integral to solving the biological diversity crisis. Data needed to achieve a comprehensive understanding of the loss of biological diversity might come from studies of:

Intensive versus extensive land use Evolution of intensive versus extensive systems; how human–land relationships affect intensity of use

Markets, laws, and traditional values How economic, legal, and social systems affect which species and ecosystems can be used and which cannot

Consequences of economic measures Effects of gross national product, discount rate, and other measures on biological diversity

Alternative systems of economic valuation Mechanisms that could address adverse effects of neoclassical and Marxian paradigms on living systems; use of common, biologically active elements (e.g., nitrogen, carbon) instead of rare, inert ones (e.g., gold, silver) in accounting systems

Socioeconomic determinants of conservation Comparing effects of different standards of living on society's conservation ethics and different conservation ethics in societies with comparable standards of living

As is the case with the other examples, data on these types of variables are uneven. Some of the independent variables consist of taxonomies of political and economic systems, such as types of markets, legal institutions, or systems of land use. Some of the dependent variables are available on an annual basis (measures of biological diversity, societal standards of living, intensity of land use) while others must be generated from cross-national surveys (comparisons of attitudes toward conservation).

Generally, the data problems raised by such hypothetical studies are the major ones facing quantitative research on the human dimensions of global change. They are, briefly:

—availability of and access to existing data;
—quality control and interpretability of data;
—lack of necessary data for parts of the world;
—inadequate time series for some variables;
—lack of measures of some key variables; and
—incompatibility and incommensurability across data sets in geographical or temporal scope.

DATA AVAILABILITY

Data are available on some social phenomena that are relevant to the human dimensions of global change: for example, trends in population growth, dispersion, and mortality, per capita energy use by nation and region, health and disease statistics, agricultural productivity for nations and regions, various indices of economic and industrial growth and decline, and political events related to the stability of governments. In fact, there is so much statistical information that it was beyond the capacity of the committee to judge its quality or its adequacy for the study of global change. Few if any scholars have a clear sense of the full range of data that are available.

Since the conclusion of World War II, many aspects of social science have been revolutionized by the increasing availability of quantitative data about social phenomena throughout the world and of computers that permit large quantities of data to be analyzed quickly and easily. Traditional administrative and census techniques have been used more intensively to gather more and more fine-grained data and more extensively throughout all areas of the world, and huge archives have been built. The sample survey, which has been greatly refined in the last 50 years, has made it possible to generalize to entire populations from a relatively small, carefully drawn sample and can therefore supplement censuses. Surveys also make it possible to gather data about issues that can be explored only through interviews.

Massive data collection efforts, beginning in the 1930s in the Western countries, have aided in the development of social theory. In some cases, data were collected because theories demonstrated that they would be useful to understanding how social phenomena interacted. National income accounts are a prime example of this interaction between the development of theory and the collection of data. In other cases, theory has allowed exploratory data analysis and the building of inductive theory.

What Is Available

The national-level data sets published by the United Nations and other international agencies are global data on human phenomena. After World War II, these agencies developed standards for gathering data, trained statisticians throughout the world, and inaugurated a vast array of serial publications of statistical data. The United Nations and most of its related agencies such as the

International Bank for Reconstruction and Development (World Bank), the International Monetary Fund, the International Labour Organisation, the Food and Agriculture Organization, and the World Health Organization have developed serial publications containing vast quantities of data. Regional organizations, such as the Organisation for Economic Co-operation and Development (OECD), have also developed and published their own statistical series.

Important series include the World Bank's *World Tables*, the UN's *Statistical Yearbooks*, *Demographic Yearbooks*, and *World Population Trends and Policies*, the OECD's *National Accounts* and *Economic Surveys* (annual reviews of member countries' economies). The United Nations Environment Programme and OECD both publish series that deal directly with environmental issues, although the data in them are much more limited than in the economic and demographic series. The quantity of data published by international institutions is huge. The World Bank's *World Tables* include data on 138 countries for a wide range of economic and social phenomena. The UN's biennial "Monitoring of Population Trends" is an excellent global synthesis of national population trends, containing demographic data on over 200 countries. Beyond being available in publications, many of these data are available in machine-readable form to individuals and organizations not affiliated with the international institutions.

In addition, institutions exist that receive, archive, and disseminate social science data. For example, the Inter-university Consortium for Political and Social Research (ICPSR), a consortium of more than 300 member universities that is administratively located at the University of Michigan, manages an archive of more than 2,000 data collections, 28,000 data files, and 190 billion characters of information. The collection includes many important surveys of individual attitudes, as well as machine-readable data sets from the U.S. Census and from international agencies. However, the collection of data sets is neither global (most of the studies are restricted to one country) nor systematic (the archive includes what was sent, rather than all data sets on any particular subject, or data sets on all the topics in a predetermined domain). Other similar archives are the University of Connecticut's Roper Center for Public Opinion Research national sample survey and the Louis Harris Data Center at the University of North Carolina, Chapel Hill.

Research on human dimensions also requires data on the nonhuman variables that affect or are affected by human action, such as trace gases and pollutant levels, land cover, precipitation and frost patterns, incidence of UV radiation, and so forth. The bulk of potentially relevant information far surpasses information on the human dimension itself. Many of the data sets are available from central sources, and there have been some efforts to build data catalogues that would allow researchers to find what they need. However, these efforts are not far along; they typically have little or no input from researchers who might use the archives to study the human dimensions of global change.

There is, however, a trade-off between using existing studies and collecting new data. On one hand, existing data are attractive because the collection costs have already been paid. On the other hand, it may well be the case that the costs of search and assembly of existing data are underestimated and their value in terms of relevance and quality overestimated. These issues must be evaluated in terms of the relative cost-effectiveness of both strategies. The time to make such assessments is *before* a decision is made to invest in large-scale data assembly or cataloguing (for extant data) or data collection (for new data) efforts. And the assessments should be based on a modest but representative sampling of available materials on key variables for pivotal nations or regions of the world. Issues surrounding such assessments, in terms of articulating criteria for judging quality and costs, are elaborated in some detail in the remaining sections of this chapter.

It is clearly necessary before all else to make existing data available to researchers so that they can find out what can be learned without expensive collection of new data. To do this, steps must be taken to establish a viable information network.

AN INFORMATION NETWORK

Because research on global change is so broad and draws on data, empirical research, and analysis from a wide variety of disciplines and countries, few social or natural scientists are familiar with more than a narrow portion of it. Giving them access to the data they need presents a major task for managing, storing, and distributing scientific information. Meanwhile, the electronic revolution has created a major upheaval in the way information is managed. The speed, density, and cost of storing, processing, and distributing information is changing more rapidly than at any time since the age of Gutenberg. However, our institutions—particularly our libraries—are adjusting to the electronic revolution much too slowly.

Putting these two points together, we believe that intensive development of electronic information management in the social, natural, and environmental sciences holds significant potential for major breakthroughs in empirical work on global change. It could both encourage global change research and help bridge the wide gaps among the disciplines engaged in global change research. By harnessing the electronic revolution, we can form "invisible colleges" of people in different locations and disciplines who work closely together.

An appropriately designed public information system would be a powerful equalizing force in the scientific community. Currently, those who are firmly ensconced in disciplines—especially those at major research universities with amply funded libraries and computer centers—have major advantages over those outside existing centers. A public information network would level the playing field for all researchers—a scientist in Oshkosh or Baton Rouge would have the same access as one in Cambridge or San Francisco. The Inter-university Consortium for Political and Social Research already plays this role for much social science research—thousands of important data sets are available to all member institutions at nominal cost.

It should be noted that a public information system would benefit the entire community of researchers on phenomena related to global environmental change. It would not be limited only to those focusing on the human dimensions of the problem or those whose work concentrates primarily on natural phenomena. To be of any use at all, the information would need to include data on natural or physical changes as recorded, for example, by remote sensing technologies, as well as changes that occur on the many dimensions of human activities that interact with the natural physical changes. (Of course, as noted above, there are huge problems of matching data collected with different conceptual frames and units of analysis.) Such an information system would contribute valuable resources for the interdisciplinary projects needed to advance the field.

The major need at this point is for governmental and private support for the necessary infrastructure for publicly shared data on demographic, economic, political, attitudinal, and natural or physical changes. This includes one-time costs, such as putting material in electronic form and building a network and archival facility to make the material accessible to researchers. The key components include facilities for storage of information; a high-speed network for exchange of information; and local facilities for receiving and processing data along with generating new knowledge and information. The National Research Council's Committee on Global Change recommended: ". . . that data be systematically archived, prepared, and disseminated at a central site and made readily accessible to interested researchers" (National Research Council, 1990b:127). While we generally support this recommendation, the committee is divided on whether or not the data and information system should be concentrated in one location. A skeletal description of the needed system might be the following:

Storage   Given the economies of scale, there should be a small number of facilities that are information libraries for large data bases, working papers, and documentation in global change. These would not need to be at one central location; specialization should be encouraged. Each center would operate as a public utility, providing services at close to marginal cost with overhead covered by federal subvention or subscriptions. Again, several existing data centers serve as models.

Data transformation   Data gathered on different temporal and spatial scales will need to be stored in a form that allows transformation to a variety of spatial or temporal formats. The conceptual and methodological issues involved in storing such data for easy transformation should be addressed. There is a substantial foundation of ongoing work in such areas as epidemiology and geographic information systems that can serve as a basis.

Cataloguing   An inventory of archived data should enable potential users to identify sources of data on the variables of interest to them, get critical information on the spatial and temporal characteristics of the data, and select data sets for their particular needs.

Transmission   The centers and major research institutions would be connected by a variety of different media. A high-speed network would be desirable for linking researchers and archives together. Telephonic communications would be a backstop for those not on a major data link or can be used for a backup. And of course, physical shipment of tapes or disks involves delays of only a few days.

Local facilities   Each research institution would decide how best to configure its local environment and would have the responsibility to determine how best to participate in data networks.

Computation   Given progress in development of work stations and in parallel processing, we believe computation is likely to be the least important part of the problem of better use of information. More important is likely to be the method of connection to the network and the cost of and access to a variety of software.

Some of the elements of such an electronic information network already exist. There is, however, much work to be done in completing the system, particularly with respect to making existing material easily available to remote users. It is noted, however, that expenditures on such a system should not jeopardize needed resources for doing the research. Any electronic data system should facilitate, rather than substitute for, the time-consuming effort of attempting to understand the data, namely, how variables are defined, measured, aggregated, commensurated, and so on. The ultimate goal of these systems is to support the process of doing research. We are especially concerned about the possible temptation to expend valuable resources on the collection of data for its own sake. Data collection and dissemination efforts must be closely coordinated with research designs and analysis plans that can guide decisions about what to collect and how it will be used in analysis.

One piece of important work for research on the human dimensions of global change involves making data from remote sensing instruments intelligible for social scientific purposes. For example, time-series observations of the land area covered by buildings may be a useful proxy for measures of population and economic activity in rural, Third World areas for which direct measures are poor. The data can certainly be updated much more frequently and at much lower cost. A careful analysis should be done of the use of remote sensing for the measurement of social phenomena. Specifically, an effort to identify existing remote-sensing information useful for human dimensions studies and to transform the relevant data into indices of human activity would be a valuable one-time investment in a global change information system.

The Earth Observing System (EOS) is a new remote sensing system for the study of global change created by he National Aeronautics and Space Administration (NASA). EOS will have important new capabilities. Social scientists should be involved in its planning to ensure that the potential of this system for studying the human dimensions of global environmental change is fully realized. In addition, many of the data requirements for research on the human dimensions of global change involve improved methods for ground observation.

We make the following recommendations:

An information network should be designed for global change research (including the human dimensions) that will make catalogues, data bases, data transformation, documentation, scientific literature, and other public information more widely and inexpensively available. On the social science side, the network should include measures of the major driving forces of global change at the lowest available level of aggregation. The decision on how to organize the network should be based on advice from a group widely representative of natural and social scientists, information systems specialists, and librarians and archivists.

As part of the network, we recommend that the federal government should seriously consider establishing a national data center on the human dimensions of global change parallel to the existing national centers for data on climate, oceans, geophysics, and space science. We are aware of the dangers of delegating responsibility for a national data center to an agency that may not be oriented to understanding the data needs of human dimensions research or to promoting a strong research effort in this area. We therefore recommend that an independent advisory committee, composed of researchers working on the human dimensions of global change and including strong representation of social scientists, be set up to oversee the work of any such federal program that may be established.

The federal government should support an effort to examine the utility of existing and forthcoming remote-sensing information as indicators of particular human activities, to validate the most promising remote-sensing indicators against ground data, and to transform the data into indicators and include them in the information network. Whenever possible, this effort should include social scientists in the process of designing and specifying the capabilities of EOS and other remote-sensing systems for measuring global change.

Social scientists should serve on the EOS Science Advisory Panel and on any advisory panels established for the EOS Data and Information System.

Costs of Data

The increasing commercialization of data is a troubling trend and a roadblock to scientific research. More and more, many of the useful data bases are being provided at extremely high cost, not only by commercial vendors (like Data Resources, Inc.) but also by quasi-governmental agencies (like the IMF or OECD). For example, a complete set of the OECD data costs tens of thousands of dollars annually—far above the marginal cost of providing the data. Clearly, pricing of major data sources to recoup costs of data collection will inhibit global change research that necessarily relies heavily on comparative international data in economics and other areas. This is an area in which the U.S. government could help by encouraging international institutions to make data available at reasonable cost to nonprofit researchers. **The committee recommends that the United States government should take steps to prevent pricing of basic data at levels far above the marginal cost of production. The government should try to influence international institutions such as the OECD to do the same.**

QUALITY AND INTERPRETABILITY OF DATA

In any research effort, careful attention must be paid to issues of data quality, and efforts to better understand the human dimensions of global change are no exception. But for several reasons the problem of data quality may be more substantial in this area than in other research efforts involving social science. First, the volume of data that will be generated is enormous. NASA estimates that the EOS will generate a terabyte of data a day when the full system, described as "Mission to Planet Earth," is in place. To illustrate the magnitude of this data stream, the daily volume is sometimes referred to as a LOC, meaning Library of Congress, because a terabyte is an amount roughly the size of the library's contents. Managing that volume of data in a way that will make it useful is an enormous problem. But in addition to the management problems, considerable effort will be required to ensure that the data being generated have reasonable levels of reliability and validity.

Second, much of the data will consist of measurements on variables that have not been given much methodological attention in the past. Considerable effort has been devoted to the measurement of fertility rates, gross national product, public opinion, etc. In contrast, relatively little attention has been given to reliable and valid ways of measuring deforestation rates, energy efficiency, social movement mobilization, and so on. Certainly, experience in developing reliable and valid measures can transfer across concepts, but applying general methodological knowledge to the specifics of global change research will require a substantial effort.

Third, many variables will be measured with new technologies. Of particular importance is the potential for making use of data from satellites and other forms of remote sensing. In the physical and biological sciences there is some tradition of using remotely sensed data, but it is a novelty to the social sciences. Very little effort has been devoted to developing methodologies that would make such data useful for understanding the human dimensions of global change. Indeed, as we note elsewhere, the social sciences, with the exception of geography, have paid little attention to space as a concept and will have to develop new theory and methodology to accommodate global change phenomena that have a strong spatial component. It is especially important that social scientists become involved in the planning of remote sensing systems and the derived data bases. If social science concerns are not introduced early on, information that could be of great value in better understanding the human dimensions of global change may not be collected and archived in a form that optimizes its utility.

Fourth, much of the data of greatest interest will have to be collected on a regional, international, or global basis. The problems of aggregation from smaller to larger units and of assessing comparability across different local or national measurements will be important. There is considerable experience in dealing with these problems in the fields of economics, demography, and comparative politics. But that experience will have to be translated so as to be useful in measuring the very different variables important to global change.

Fifth, the conceptualization and measurement of many key variables must be interdisciplinary. In most cases sound conceptualization and measurement must span several social science disciplines; in many cases it will involve the physical and biological sciences as well. For example, a land use taxonomy must be sensitive to the biological and physical characteristics of a particular land use, such as species diversity that typically accompanies it, its albedo, and its percolation rate. But the taxonomy must also be sensitive to the economic function of the land use, its political regulation, and its cultural meaning. It may be that several taxonomies are required, but if so they must be coordinated to produce useful data.

Finally, although quality and interpretability is of the utmost importance for research on the human dimensions, attention must also be given to issues of cost-effectiveness. New large-scale data collections are costly enterprises. Just as some research projects would be given priority over others, so too would some variables take precedence in data collection efforts. It will be necessary to develop criteria for determining where to place investments in data-collection efforts. In part, this is a problem of developing an accounting system similar to that described in Chapter 3. It is also a problem of estimating the quality of data available on the high-priority variables. For these reasons, it is important to include, as part of the assessment of quality, the likely costs involved in obtaining the data. Importance, quality, availability, and cost-effectiveness are all criteria for judging the feasibility of mounting large data-collection efforts.

PROBLEMS OF RELIABILITY AND VALIDITY

Threats to validity and reliability are not unique to data dealing with global change, so we can draw on the more general methodological literature on techniques to improve measurement. In some cases that literature is quite substantial; in other cases it is not. The issues noted above suggest that existing methods can be a source of valuable insights but will require some translation and modification as they are applied to global change. With this is mind, we believe that there are four major issues that should be given special attention.

First, the raw numbers used to generate a measurement may themselves be inaccurate because of errors in survey responses, official records, and so on. In some cases these errors are random and simply reduce reliability; in others they are systematic and thus introduce bias and reduce validity. The errors may be caused by the happenstance of human error and imperfect memory or by intentional obfuscation and deception. Decades of experience with official statistics, survey procedures, censuses, and so on have generated a rich repertoire of methodological tools for identifying and minimizing these sources of error (Bollen, 1989).

Second, there are problems of sampling and coverage. Much of statistical analysis is directed toward understanding sampling error, but nonreporting and other sources of bias remain a very difficult problem. The biases introduced by poor coverage are likely to be substantial in research on global change. It is the affluent industrialized nations that generally have the most sophisticated statistical systems and the richest data bases, while the poorer nations often have no estimates for many key variables. For example, estimated annual emissions of nitrogen oxides are available for 19 industrial nations, but for only 3 developing nations (World Resources Institute, 1990: Table 24.6). The same problem applies within a nation or region. The middle class and the urban are more likely to be included than the poor or rural in surveys and other data collection efforts. In the 1950s and 1960s, when concern with population growth was at the center of the international research agenda, the problems of coverage for basic demographic data such as fertility rates and the prevalence of contraception were similar to those that exist now for key global change variables. Careful efforts over several decades have greatly improved coverage as well as the quality of measurement. We hope a similar effort can be undertaken to improve understanding of the human dimensions of global change.

A related issue arises when it is necessary to estimate values on key variables for some countries or regions. Estimation may be required because of a lack of coverage. It may also be necessary because the reported data are suspect. In many cases, data are supplied by a local or national body that may have a strong political interest in the values reported and may distort measurements either intentionally or by using methods that are biased. In other cases, the lack of infrastructure may prevent the local or national body from providing reliable and valid data. Whatever the reason, if the reported data are badly flawed, it is advisable to use multiple methods in order to cross-check results. Care must be taken to carefully evaluate these estimation procedures. Again, the history of demography and economics since World War II suggests that it is possible to develop sound estimation and validation procedures to supplement official reports and primary data. Of course, the ability to develop estimates and checks depends on the character of the variables being estimated. For many important demographic and economic variables, the logic of accounting and internal consistency can be applied to estimation. This should also apply to some of the key variables in global change.

A fourth problem arises because different jurisdictions may use different definitions and thus produce noncomparable data. Variables must be defined taking account of theoretical and practical considerations, and both theory and practical context will vary from place to place. Again, demographers and economists have considerable experience dealing with problems of comparability.

RESEARCH ON MEASUREMENT PROBLEMS

We begin by raising the general point that methodology can be improved by careful, focused research. We suggest that projects intended to enhance knowledge of the human dimensions of global change should be assessed not only on the quality of the methodology used but also on their likely contribution to advancing methodology. We also suggest that some projects that are intended primarily to enhance methodology should be supported.

For many variables important to the study of global change, there is a long chain of aggregation and integration that leads from data collected in the field to published reports or data bases. The procedures used at any step of the process may degrade reliability and introduce bias, but those problems may not be apparent in the final data table. It is easy to find published values of gross national product per capita, rate of population growth, or CO_2 emissions for most nations of the world, but it is more difficult to document the full chain of research that produces those numbers for each nation. Much cross-national research simply accepts the published figures at face value, ignoring serious problems that may be hidden in their production. A few case studies that trace the production of data on key variables would be very helpful. Much methodological information seems to be available only in the minds of official statisticians or in a fugitive literature of internal documents, manuals, and memos. By examining the production process, we will develop a better sense of potential threats to validity and reliability and thus will be in a better position to recommend improved methodology and offer appropriate cautions regarding the use of existing data. (See Cook and Campbell, 1979, for a checklist of common threats to validity, including measurement and data analysis problems.)

Modern methodology also moves beyond simple dichotomies of data quality such as "good" and "poor" or "reliable" and "unreliable" to procedures for both identifying the reliability of indicators and for aggregating them into more reliable indices by taking explicit account of measurement error. For example, Bollen's (1980) work on political democracy has both identified the character of errors in existing measures and developed a more reliable indicator that has seen extensive subsequent use. There are long traditions in psychology (measurement theory), sociology (path analysis), and econometrics (errors in variables procedures) addressing these issues. In the last decade they have been unified into a powerful set of techniques that can advance our ability to make appropriate use of flawed data (Bollen, 1989). Again, we would suggest that a few targeted studies using these techniques to analyze multiple indicators would greatly enhance our understanding of data quality and suggest methods for improving both conceptualization and measurement.

Problems of limited validity and reliability in the measurement of key variables are not unique to the social sciences. Physical and biological science data sets of the sort used for the study of global change are fraught with the same problems that have had the attention of social scientists for decades. Social scientists have developed many procedures for dealing with such data and for improving data quality. Analyses and data collection efforts in the physical and biological sciences could benefit from application of some of them. In particular, the statistical procedures developed for multiple indicators, for analysis in the presence of missing data, and for analysis in the presence of unusual but possibly valid observations could easily be transferred to the physical and biological sciences. So could some of the technologies for improving the reliability and validity of data collection.

DATA COLLECTION NEEDS

To what extent do existing raw data suffice for the needs of global change research? Data needs tend to be very researcher-specific, so one person's experience may be quite different from another's. On the whole, data relating to many critical variables concerning global change exist in some form or another, although (as already noted) the data are sometimes of very poor quality and often of unknown quality.

Data that are not currently available will undoubtedly be required for the study of the human dimensions of global environmental change. What they are will become clearer as the inventory of existing data is completed and as the studies proceed. Even at this early stage, however, it is apparent that there are significant lacunae. Environmental quality, valuation of nonmarket activities, and leisure-time budgets have been areas in which data are poor relative to the plentiful data available for the study of financial markets or labor markets. Data on the land market, however, are very sparse. Broad-based analyses of land values, which are needed for the study of the human impact of a rise in sea level, require such data. Baseline data on public and elite attitudes toward global change on a worldwide comparable basis and on the enactment and implementation of environmental policies also do not exist.

Survey data in particular tend to be sparse, partly because surveys have been expensive and highly vulnerable to the budgeteer's ax. Improved survey data in the areas of energy use, perceived environmental quality, and recreation and time-use patterns could help improve our understanding. In addition, surveys in developing and socialist countries are particularly important for comparison with the responses of individuals in developed capitalist countries.

Undoubtedly, additional kinds of new data will have to be gathered. It is important, however, that this be done carefully, deliberately, and generally in the context of actual studies. There are important synergies between the development of theory and the collection of data, so premature data collection may yield data of little value for future research.

Since the study of the human dimensions of global environmental change aims at understanding dynamic processes, it is crucial that baseline data be available so that change can be observed. Special priority should be devoted to the collection of baseline data on key measures, including indicators of the major driving forces of global change for which data do not already exist.

Governmental and intergovernmental agencies will be collecting the bulk of social data, as they do now. They may also be developing new indices, for instance, supplements to national accounts for the costs of resource use and depletion and of environmental services and degradation. In both these roles, it will be important to ensure that the new data fit easily with existing time-series data. It is therefore important that the scholarly and governmental communities work together constructively. To more carefully assess needs for new data, **we recommend that an inventory of data and survey needs in selected areas be developed and examined to determine whether expanded data collections in these areas are needed.** The areas should include the following, among others:

—land use and food production
—economic activity
—consumption of energy and materials
—human health
—population trends
—environmental quality
—environmental attitudes

The inventory should be developed through consultation among social and natural scientists, governmental and nongovernmental statistical agencies, and data base management and archiving specialists associated with the global change data network. It should be updated periodically to take into account new data needs uncovered by recent research.

This project would benefit from oversight provided by an interdisciplinary advisory group consisting of leading researchers as well as representatives of the key agency sponsors of research on the human dimensions of global change. The advisory group would be responsible for planning the project's design and implementation.

ANALYTICAL DATA AND ACCOUNTING

Studies of global change have been hampered by a number of problems with analytical data, that is, social indices or aggregates that take raw data and put it in a form useful for analysts. Examples include data on gross national product, indices of inequality of land or income distribution, average energy prices, and so forth. The major issues have been inadequate data outside the advanced countries and excessively narrow accounting systems for all countries.

It is well known that reliable social and environmental data have been sparse outside the OECD countries. This difficulty may be of little consequence for national social, economic, or environmental policies, but it is significant for analyses at the global level. Because poor countries often do have not the resources or statistical personnel to collect and process data, their social, demographic, economic, and environmental data are usually of poor quality. Often the best data on some environmental variables come from satellite studies. In addition, the poor quality of data from socialist countries has been compounded by antiquated accounting standards and politically motivated distortion of data. A major effort to improve the accounts of socialist and low-income countries would clearly benefit both the countries and global change researchers.

Existing systems of national income accounts omit many sectors that are important for global change research. Three critical omissions are nonmarket use of time (which is important, for example, where climate change affects recreation); use of depletable natural resources (such as oil and gas); and environmental spillovers and abatement activities (such as the costs of oil spills or the changes in amenity values of unpriced resources). Research has moved ahead sporadically on some of these topics over the past 20 years, although little progress has been made in the United States. Researchers in Germany, Japan, and other countries have made major contributions to augmenting the national accounts, while some attempts have been made to see how such changes might affect the accounts of a developing country such as Indonesia. This work needs to be undertaken either through developing augmented accounts in the Commerce Department or by supporting research in this area. In the area of analytical data, we make two recommendations: **The United States should support major improvements in data, accounting practices, and national accounts for developing countries and current or former socialist countries.**

The United States should develop augmented national accounts to include nonmarket productive activity, depletion of natural resources, and the value of environmental spillovers and resources. These factors would contribute to more accurate indicators of the gross national product.

SAMPLING NATIONS

Research on the human dimensions of global change poses an analytical dilemma. On one hand, understanding of causes, responses, and consequences depends on analyses of data collected at a global level of analysis. On the other hand, any attempt to assemble data for all the nations of the world would be prohibitively expensive and impractical since data of high quality probably exist only for a small set of nations. What is needed is an analytical strategy that would serve to economize collection, dissemination, and analysis while not forfeiting our ability to derive implications for the globe. One approach is to develop a systematic sampling design for national-level data collections.

An appropriate sampling of nations would include the large and pivotal environmental countries such as the United States, the Soviet Union, China, India, and Brazil; two or three countries from Europe; and the remaining countries selected from a pool of nations for which data resources have been found to be adequate for analysis. This approach entails an evaluation of resources (and costs) prior to selecting countries for inclusion in the sample. For the countries chosen, there would be national-level data collections bearing on the relevant institutional structures, assessment of environmental variables beyond what can be inferred from satellite data, trends in public opinion, and others as illustrated by the lists presented earlier in this chapter. It is true, however, that the nation is not the only relevant sampling unit. Other territorial sampling units may be appropriate, including regions and cultures that cross the borders of recognized polities. Any sampling frame could be stratified on the basis of classifications drawn from satellite measurement as well as types of social and economic institutions, land tenure patterns, population densities, level of technological development, and so on. (A variety of alternative sampling designs are presented in Kish, 1965, and in the earlier work by Hansen et al., 1953; these sources remain the classic treatments of this subject.)

Cross-national data collections and analyses are at an intermediate level, between global-level data and intensive local studies. Operating at this level of analysis has the advantage of providing greater richness of information than what is likely to be available at a global level while affording better coverage of key variables than what is usually available from a scattering of episodic case studies. It is also a level at which such critical processes as policy formulation and implementation, discussed in Chapter 4, occur. Therefore, we suggest that research on the human dimensions of global change utilize sampling strategies that would take advantage of the richness of available information without forfeiting coverage.

INTENSIVE LOCAL DATA COLLECTION

As we have noted in Chapters 3 and 4, data analysis at the global and national levels can yield only part of the necessary understanding. Important human causes of and responses to global change occur at the local and individual levels and require study at those levels. To relate activities at those levels to their global effects requires understanding of how and why human activities and responses differ from one locale to another; this understanding, in turn, requires detailed knowledge of the locales. These considerations provide a rationale for the intensive study of a relatively small number of locales, selected so that they can be compared with each other to yield the desired knowledge.

The identification and study of locales for intensive study requires rather detailed information and assessment about the relationships between environmental and social variables. Typically, the larger the spatial and temporal scale of analysis, the more difficult to achieve and manage the requisite level of detail; yet the smaller the scale, the less significant the case in absolute terms and the less transferrable the results to other situations. Studies in locales, despite their limited generality, can offer valuable insight into comparative human–environmental relations if they are carefully selected. They can generate more complex hypotheses by testing the initial assumption that similar environmental and socioeconomic circumstances generate similar patterns of environmental change and of response to change. To identify such similarities, detailed analysis will require common protocols that allow for comparability among circumstances. Studies of locales can also help build capabilities to monitor at the local level environmental changes and their consequences that are also of interest at the global level. And they can be used to develop measures that can then be applied in other areas of the world.

Analysis of local causes or responses to environmental change is complicated by the interconnections between locales. The importance of physical impacts received from other regions (e.g., acid rain) or from the global system (e.g., global climatic change) and of driving forces originating elsewhere, such as commodity demand, means that analysis cannot be successfully conducted only within the bounds of the region in question. Locales that import resources from elsewhere are resistant to local environmental changes but vulnerable to changes occurring in the supply areas. Analysis must also address these broader spatial linkages, the difficulties of doing so notwithstanding.

These considerations complicate but do not reduce the importance of the intensive study of locales. Such studies are worthwhile not only for the substantive reasons noted, but also because they have the potential to bring together interdisciplinary research groups in continuing interaction—a much-needed attribute of effective global change research, as we noted in Chapters 3 and 4.

A similar idea is put forth in the recent report of the National Research Council's Committee on Global Change, *Research Strategies for the U.S. Global Change Research Program*. They suggest that regional resource sites be established "where relevant studies from the natural and human sciences can be conducted in concert, and their data sets pooled" (National Research Council, 1990c:122). This approach would insure against a lack of coordination characteristic of projects in which, for example, "ecological data are collected at one site, tropospheric chemistry data at another, and human activity data at a third" (p. 122).

But how should locales be selected for intensive study? Because of the importance of comparative research on locales, a haphazard process would be a terrible waste of the potential for knowledge.

One strategy for selecting locales for detailed study is to choose *critical zones*. By this term, we mean locales in which human-induced degradation in the physical or biological conditions of the area are contributing (or are expected soon to contribute) to a loss of things the occupants value (for alternative definitions, see Price, 1989). Such zones are interesting as microcosms of some possible undesirable futures of global environmental change. The study of such areas would allow determination of the factors contributing to environmental changes of particular interest and to the various responses to such changes at the local level. It would also generate more rapid results than studies of slower-changing or larger areas, because with more rapid change, relationships are more apparent. It would be appropriate to begin with relatively modest pilot projects that alert investigators to possible problems that can be dealt with prior to developing the infrastructures needed for the research on patterns of social, economic, and ecological change in the chosen locales (see also National Research Council, 1990b).

It would also be possible to conduct intensive studies of *sustainable zones*, that is, areas characterized by presumably sustainable use of natural resources. Such studies would best be comparative between similar or nearby regions with different patterns of resource use, one presumably more sustainable than the other. The logic is similar to that for critical zones.

Other criteria for selecting locales might also be defensible. **We recommend that intensive study of locales should be included in the global change research program**. The locales and the programs to study them be selected according to criteria including:

—groups of locales should be chosen together on grounds of similarity and difference that would illuminate important global change questions (e.g., similar natural environment but different political systems);

—social scientists and natural scientists should work together on the same ongoing research;

—projects should preferably contribute new methods or measures that could be applied elsewhere; and

—projects should employ measures for the locales taken from global data archives and return quantitative data on the locales to the same archives.

It should be noted, however, that the suggested model for doing such research is collaborative field work in the anthropological tradition: researchers would remain at the site only until completion of the project; they would not establish a center at the field site. We are *not* recommending that permanent institutions be developed as, for example, in the mold of the biological field station.

7

Human Resources and Organizational Structures

The previous chapters make evident the complex conceptual and methodological problems faced by researchers studying the human dimensions of global change. Progress in this field will depend on laying a foundation for the necessary long-term research effort, that is, on how well we are able to organize human resources for the study of global change. The problems are fundamentally interdisciplinary and require an understanding of diverse perspectives and methods that transcend the traditional disciplines: research agendas must incorporate perspectives from the social and natural sciences, as well as from humanist scholars. In many cases, individuals will be required to reorient their careers to make global change a major object of study. Just as individual scholars may be required to change their research orientation to address global change issues, new arrangements are also needed in funding agencies and research institutions. This chapter considers ways to organize large-scale research efforts to incorporate the needed cross-disciplinary skills and perspectives. It addresses the institutional infrastructure for research, training programs for scholars and scientists, and the organization of specific research agendas. We believe that careful attention to these issues is critical to the success of the research program on human dimensions of global change.

INSTITUTIONAL STRUCTURES

We begin with two key questions: Will major research institutions here and abroad reorient to focus their resources on global change research? Can such research flow directly out of the mainstream concerns of existing academic disciplines, or does it require fundamental restructuring?

These questions offer a starting point for discussions of how human and capital resources should be allocated to yield practical and theoretical knowledge about global change. If it is possible to reorient institutions and disciplines to make global change more central to existing research agendas, then how do we invest resources to achieve such a reorientation in the most efficient and productive ways? If, however, existing institutions and disciplines seem unlikely to respond adequately to global change as an issue, then what alternative institutional structures and disciplinary loci are likely to yield the results we seek?

A sudden infusion of money in large, temporary doses rarely produces sustained, effective programs. Existing disciplines and institutions either take advantage of the new money to serve related research agendas, or come to depend on external funding and do not incorporate the new research agenda into their *internal* financial and intellectual accounting. There is a risk that research on global change will follow this pattern. In this situation, institutions funding research have essentially two options: (1) to recognize from the outset that their goals are short term and hence may not make a lasting impact on disciplines and research institutions or (2) to commit significant resources to strategic planning and the development of long-term institutional resources. With respect to global change research, the most successful plan is likely to be the one that incorporates both these options.

The "war on cancer" and other major initiatives in biomedical research offer examples of strategic decisions to build specialized multidisciplinary research resources to sustain a field of priority research. Existing disciplines and institutions should be encouraged to investigate global change, and a limited set of new institutional structures should be developed in areas in which existing resources seem inadequate. It is essential that funding for global change make a distinction between short-term investments in existing research programs and long-term investments that are designed to build new research institutions. Both are important, but the second requires sustained, long-term support—for clerical staff, office space and equipment, travel funds, and the availability of postdoctoral fellows.

Interdisciplinary programs are almost always at a disadvantage in research institutions, particularly universities. Although they often begin with much fanfare and develop a large following among students, they rarely enjoy the institutional support given to standard academic disciplines. Individual scholars who commit time to such programs often do so to the detriment of their own careers. To promote a lasting institutional presence for a new interdisciplinary research program such as human dimensions of global change, funders should create explicit incentives to encourage an identity for the program within its larger institutional setting. In a university, a healthy program will eventually control its own faculty appointments and often key resources such as budgets and space, in order both to support new researchers and to attract the interest and resources of individuals who are already present but not yet committed to global change as a research agenda.

One viable model is to create university-wide committees on global change, endowed with critical resources but not usurping the traditional roles of departments in hiring and granting tenure and degrees. (The Committee on Population Biology at the University of Chicago, formed in the early 1970s, operated in this manner.) The loss in autonomy suffered by this arrangement is traded for a gain in access to productive researchers in a variety of natural and social science disciplines. To hold the interest of these faculty members, the committee can use its resources to give released time, visiting appointments, and fellowships; sponsor research start-ups; and hold regular colloquia.

Planners should remember that the 1990s are likely to be a highly competitive period among university faculties and research scientists. Institutions faced with widespread retirements among their senior scholars and scientists will find themselves trying to hire from a pool of younger people that is smaller than the demand for their talents. At the same time, universities in particular are experiencing serious pressure on their traditional income streams just at the moment that their expenditures on faculty salaries seem likely to rise. Under these circumstances, agencies promoting research on global change have an opportunity to encourage institutions to hire in this area. Chaired professorships, salary support, and research funding as well as more modest investments, such as underwriting special journal issues focusing on global and/or environmental change, special conferences, workshops, seminars, and lecture series are important in establishing the field. Institutionally based funding for global change research by scholars in midcareer seems a particularly fruitful way to leverage resources. **The committee recommends that potential sponsors of global change research be attentive to the problems of interdisciplinary research in most institutional settings and address some of their support to building institutional bases for research.**

NATIONAL CENTERS FOR RESEARCH

Well-funded national research centers have played a major role in advancing both basic and applied research in a number of areas within the social sciences. Notable examples are centers for research on population issues, on the analysis and resolution of social conflicts, and on international studies. Centers on these topics have served to bring together groups of scientists and to promote communication among scholars that would not occur if they were operating within traditional discipline-based departments. Centers have also served to overcome some of the institutional impediments to progress in interdisciplinary fields of study discussed above.

It is our judgment that the field of the human dimensions of global change, like these other fields, requires interdisciplinary communication beyond what typically occurs in universities. Based on the experience gained from centers in those fields, the committee concludes that the human dimensions of global change is an emerging field of inquiry that is ripe for the same sort of treatment. **We recommend that about five national centers for research in this area be established in the immediate future, and that these centers be truly interdisciplinary, combining the social and natural sciences.** The centers should be supported by a consortium of government and private sources that will make a long-term funding commitment. Long-term survival of these centers would be enhanced by weaving them into the fabric of the institutions of which they are a part. Endowed chairs and split tenured positions between the center and appropriate academic departments would contribute to the stability and status of centers based in universities.

In establishing these centers, several models are worth considering. One is to create interdisciplinary research centers within the institutional structures of universities, such as the Office of Population Research at Princeton University. Another model is to create semiautonomous centers that are loosely affiliated with universities, such as the Scripps Oceanographic Institution. A third model is physically separate centers operated by consortiums of universities, such as the National Center for Atmospheric Research. A fourth model is the completely independent center devoted to research on environmental issues, such as Resources for the Future. Yet another possibility would be the sorts of government-operated centers that conduct both intramural and extramural programs, like those managed by the National Institutes of Health.

If it were necessary to choose among these models, the committee judges that the best case could be made for maintaining close links between the centers and universities. This judgment is based on the observation that most of the significant advances in our understanding of human behavior have been made by investigators in university settings. A useful model is provided by demography. Since World War II, funding from public and private sources, much of it along the lines we have outlined, has allowed this field to advance rapidly and become a respected and highly interdisciplinary concentration within the social sciences. Knowledge of the causes and effects of population growth has increased substantially, and much of that knowledge has been translated into applications.

Independent centers have shown a tendency to focus their energies more on applied research and, in some cases, to become associated with one or another perspective. It is noted, however, that there are topics relating to the human dimensions of global change that are most appropriately conceived at the intersection between basic and applied research. These problems are perhaps most appropriately investigated at government-operated research centers, like the widely proposed National Institutes of the Environment, or at freestanding research think tanks, such as Resources for the Future.

Regardless of their location, centers should be established for the long term. Sponsors should therefore look for evidence that institutions seeking to establish centers are making a long-term commitment to them. The centers need to be rooted in environmental social science but also maintain a commitment to collaborative work with natural scientists.

TRAINING

Funding agencies have long sought to encourage research in fields with important policy implications by supporting research by graduate students, and we assume that global change research will follow a similar path. Graduate and postdoctoral fellowships are an obvious place to target global change budgets, being especially attractive because of their relatively low cost and relatively high likelihood of inducing young people to make careers for themselves in this area. Given the special challenges of global change research, such fellowships might usefully be supplemented with travel funds that would enable young scholars and scientists to do work in distant locations they might not otherwise be able to reach. The fellowships should be available to graduate students, postdoctoral students, and mid-career scientists on a competitive basis. While fellows should be allowed considerable freedom to design their own programs, they may choose to affiliate with one of the national centers for research discussed above. **The committee recommends that new programs of graduate and postdoctoral fellowships be established to support global change research. In addition to basic funding, generous grants for travel should be provided**.

There are additional issues relating to training. One is that the study of global change requires communication and cooperation among scholars and scientists in many traditional academic disciplines. Such cooperation is less frequent in university settings than one would like, suggesting the need to explicitly foster opportunities for cross-disciplinary synthesis. One possible device is to arrange for annual meetings of all graduate students and postdoctoral fellows who are working on different aspects of global change. Such gatherings should be long enough—no less than a week and preferably two weeks—to build a genuine sense of community and collegiality among their participants. Annual conferences would represent a lost opportunity if geologists talked only to geologists, ecologists to ecologists, economists to economists, and so on—yet this is exactly what will happen unless conference planners ensure that individuals from different disciplines plan projects requiring their active joint participation. Meetings should be designed to mix individuals from different disciplinary backgrounds to promote precisely the kinds of intellectual cross-fertilization that often do not occur in more traditional settings. Since much global change research is likely to be collective, requiring the talents and perspectives of individuals from many different disciplines, conferences should be organized to encourage group interaction of just this sort so that students become accustomed to it early in their careers. Analogous conferences for scholars and scientists already in mid-career are also desirable.

In seeking to invest wisely in support of global change research, attention to the likely career trajectories of individuals who will do research in this area is important. Most scholars and scientists will take the risk of investing their time and talents to study global change only if long-term career development and stability seem possible. One idea is to establish a "transportable" 5-year package of dissertation support, postgraduate salary, and research funding. This opportunity could provide an incentive for students facing the choice of thesis topics. Being awarded one of these prestigious pre- and postdoctoral fellowships would make a graduate student attractive to a potential employer upon graduation. It is likely that many universities would be interested in recruiting junior faculty with a 5-year support package.

These questions point to a special and paradoxical challenge for funders who wish to promote research in an interdisciplinary field like global change. Although the purpose of such funding is to encourage scholars and scientists to move beyond and even transcend existing disciplines, to be effective in the real world, research must also flow out of and speak to the special concerns of those disciplines. Unless global change emerges as a new specialty in its own right, which seems unlikely, the people who study it are likely to retain their principal intellectual identities as geologists, economists, climatologists, historians, ecologists, political scientists, and so on. They thus face the classic problem of designing research agendas that address interdisciplinary questions while yielding results that will speak to disciplinary interests. Graduate students are particularly vulnerable in this regard. Their dissertations will do them little good in practical terms if they fail to ground their interdisciplinary work in a disciplinary framework that potential employers—academic departments and other traditional discipline-based institutions—will recognize. To fulfill its own goals in a practical and humane way, any program of graduate funding on global change needs to address this problem intelligently.

These considerations have implications for the development of criteria for evaluating research proposals, as discussed in earlier chapters of this report. Proposals should be evaluated according to their ability to synthesize interdisciplinary questions about global change with the best theoretical and programmatic work in the relevant disciplines. The idea here is to reward proposals that are both theoretical and problem oriented, maintaining links to the disciplines without forfeiting a problem focus. Examples of attractive candidate topics are decision making under uncertainty, collective-action processes, social conflict, and the role of institutions as determinants of collective outcomes. These criteria are especially important to use in evaluating grants in support of graduate and postdoctoral research in this area. Program directors should be attentive to the implications of global change research for the long-term career prospects of younger scholars within their respective disciplines. Failing to do so may frustrate the ultimate goals of supporting global change research.

INSTITUTIONALIZING COOPERATIVE RESEARCH

Major research projects on global change should incorporate the perspectives of both the natural and the social sciences, to say nothing of the humanities. Funders should look askance at any research proposal on the human consequences of global change that fails to include scholars with suitable expertise in social science, and we trust they will be similarly dubious about proposals that address human consequences without some grounding in the natural sciences. Much might be gained, for instance, if a research team examining the socioeconomic consequences of global change were encouraged to have a climatologist and an ecologist among their number; conversely, there would be similar gains if geological studies of coastal flooding were encouraged to have an economist and/or sociologist aboard. Since natural and social scientists are not generally accustomed to working together, funders can create strong incentives for doing so. If they fail to do so, chances are that critical interdisciplinary questions simply will not be asked. Sponsors of research on global change are in a position to influence the way that research is organized. By controlling resources needed for interdisciplinary projects, funders can suggest that major investigations be organized as teams that include relevant social scientists and humanists, in addition to natural scientists. We believe that any global change research proposal that includes only natural scientists or only social scientists should justify the failure to include personnel with relevant expertise. Funders can use their leverage in these ways to promote appropriate interdisciplinary collaborations. We are not suggesting that funding agencies regulate the composition of research teams, nor are we calling for a quota system for disciplines. Rather, we suggest that agencies offer incentives to investigators who form multidisciplinary research teams when it is appropriate to do so.

Here again we encounter the paradox of interdisciplinary questions having to yield disciplinary answers. From the point of view of an economist joining a research team studying ozone depletion, or an ecologist joining a team studying climatic threats to temperate agriculture, his or her own work must yield published results that other economists and ecologists will recognize as significant and therefore worthy of career advancement. As long as global change researchers retain their principal identities in the existing disciplines, and as long as career rewards flow through those disciplines, large-scale research projects should yield significant interdisciplinary results that also advance the intellectual agendas of the participating disciplines. In practical terms, this means that research projects will ideally yield major reports on global change as well as individual monographs on its economic or ecological or other disciplinary aspects. To solve this problem, we must enlist the support of mainstream academic disciplines to define the ways in which global change research can speak to the core disciplinary questions of relevant fields. Possible ways of doing this include special conferences organized within a targeted discipline but including participation by outside scholars, special issues of leading disciplinary journals devoted to discussing global change as a research area, and discussions within the major scholarly and scientific associations about ways in which they might put global change onto their institutional agendas.

In our judgment, the social science professional associations can play significant roles in this area. Among other things, they can allocate space on the programs of their annual conventions to panels on the human dimensions of global change, organize conferences designed to explore disciplinary contributions to the study of global change, arrange for special issues devoted to global change in the journals they sponsor, include information in their newsletters about global change research activities, and provide advice to association members interested in identifying ways to become involved in global change research. A necessary condition for action on the part of the professional associations in this realm is the availability of material resources. **We therefore recommend the initiation of a program of competitive grants to provide associations with incentives to become active in this field.** One objective of these activities is to develop frameworks that can guide the design of interdisciplinary research projects that will be conducted with enthusiasm by scholars and scientists from the individual disciplines. A start in the direction of providing an interdisciplinary framework is made in the earlier chapters of this report.

The review panel process of major grant-giving agencies may easily frustrate interdisciplinary research. Grant-giving programs are usually organized along disciplinary lines, so that review panelists are generally drawn from particular disciplines with their own special evaluative criteria and research agendas. Members of one discipline often have little understanding of the theoretical orientations and methodologies utilized by members of other disciplines. Thus, disciplinary perspectives can play a counterproductive role in determining whether research receives funding. Programs that fund global change research must therefore be especially careful to draw their panelists from a wide array of disciplines and to make clear their expectations that disciplinary criteria should not bias evaluations. A two-step review process in which disciplinary panels evaluate the particular disciplinary contributions of a major collective project, and interdisciplinary panels evaluate its overall significance as a contribution to understanding of global change may be desirable in some instances.

ORGANIZATIONAL BARRIERS TO RESEARCH
IN THE FEDERAL GOVERNMENT

So far, this chapter has addressed issues relating to researchers and research institutions. The evolution of global change research in the federal government has also erected serious barriers to the support of research in human dimensions of global change. The U.S. government, which supports research on human interactions as part of the U.S. Global Change Research Program (USGCRP), is simply not structured appropriately for managing the human interactions science element. There is an almost complete mismatch between the roster of federal agencies that support research on global change and the roster of agencies with strong capabilities in social science. Of the agencies supporting global change research, only the National Science Foundation (NSF) has a strong social science capability—and NSF accounts for only a small portion of the budget for human interactions research. The major sources of support are mission agencies, including the Environmental Protection Agency (EPA), the National Aeronautics and Space Administration (NASA), the National Oceanographic and Atmospheric Administration (NOAA), and the departments of Energy, Interior, and Agriculture, none of which has more than limited expertise in a few fields of social science. Historically, the federal agencies with missions in global change have not considered environmental social science to be a significant part of those missions. And except for NSF, the major agencies concerned with understanding human systems (e.g., the National Institutes of

Health, the Census Bureau, and the departments of Health and Human Services, Housing and Urban Development, Labor, and Education) have neither major environmental missions nor any role in USGCRP. Because of this division of labor, no entity in the federal government has both the responsibility and the human resources to develop and update a comprehensive research agenda for human interactions.

The problem is not easily solved. Increased control over the human interactions science element by social science agencies would not work. Added NSF involvement might strengthen basic environmental social science but would probably not lead to development of major areas of applied research. Social mission agencies could contribute in a few areas, such as by managing studies of the consequences of global change for health and the delivery of human services or developing educational strategies to support mitigation or adaptation. But many of the most important practical questions concern energy use, land use, environmental management, adoption of new technologies, and use of technical information and so are in the domain of mission agencies such as the Department of Energy, EPA, USDA, and NASA.

The technical mission agencies, however, are not currently organized to manage environmental social science research. Social science has never been a central part of their missions, and it may not have proven particularly useful to date in addressing applied problems considered most critical to the agencies' concerns. This is not to suggest fault on the part of either the social scientists or the agency policy makers: to be relevant requires a dialogue between these communities that has not so far occurred.

The federal government should develop a strategy to ensure that the human interactions research agenda is designed and administered by organizations committed to understanding both environmental and human systems. This implies either changes in the orientation and staffing of critical mission agencies or considering new institutional approaches. The Committee on Earth and Environmental Sciences (CEES) might delegate important human–environment issues to NSF, or to particular mission agencies with the requirement that they bring on new staff or outside expertise to help with research agenda setting and management. **We recommend that mission agencies that usually support only applied research in the social sciences but have basic research programs in the natural sciences dealing with global change initiate support of basic research in the social sciences targeted on specific topics relating to global change.** This support should be managed and guided by social scientists added to agency staffs for the purpose, outside advisory groups, or both. Where such approaches do not seem likely to be sufficient, government might also create new organizational entities that bring social and natural scientists together around focused human–environment issues, both basic and applied.

CONCLUSION

The necessary advances in theory, methodology, and data discussed in the earlier chapters will not be made without taking into account the concerns about human resources and organizations discussed in this chapter. In particular, we need to foster stronger partnerships between social and natural scientists and to stimulate social scientists to transcend the perspectives and methods that guide disciplinary research. These goals will not be achieved without a long-term commitment to providing adequate funds and appropriate institutional structures to administer these funds and sustain participation by a sufficient number of well-trained scientists.

8

A National Research Program on the Human Dimensions of Global Change

Efforts of this scope and magnitude invariably yield numerous conclusions, many of which lead without difficulty to recommendations. This book is no exception. In the interests of establishing priorities and achieving clarity of exposition, we have chosen to present many of our recommendations in the sections of the report to which they pertain. In this concluding chapter, we lay out the principal elements of a comprehensive national research program on the human dimensions of global change. It is based on the analysis presented in this report and can, in our judgment, be put in place during the next 3 to 5 years.

The program plan we recommend consists of a package of five major elements: (1) an enlarged program of investigator-initiated research on the human dimensions of global change, (2) a program of research targeted or focused on selected topics relating to the human dimensions of global change, (3) an ongoing federal program for obtaining and disseminating relevant data, (4) a broad-gauged program of fellowships to expand the pool of talented scientists working in this field, and (5) a network of national centers dedicated to the conduct of research on the human dimensions of global change. Each element of this program plan is action oriented and directed toward the National Science Foundation (NSF), other appropriate federal agencies, and private funding sources. Each has implications for funding that we address in the final section of this chapter.

OVERARCHING ISSUES

The social and behavioral sciences have a vital contribution to make to any program aimed at enhancing our understanding of global environmental change. The global changes of interest today differ from those of the past precisely because they are products of human activities that have been accelerating rapidly in recent times and because the changes themselves are occurring at a pace likely to call for clear-cut responses within a single human lifetime. It follows that we cannot hope to understand the causes of these global changes or devise appropriate responses to them in the absence of adequate knowledge about the human dimensions of global change. So far, however, only modest efforts have been made to integrate the social sciences into global change research programs in the United States or elsewhere. We have traced this situation to a number of interactive factors.

The search for enhanced understanding of global environmental change requires, to begin with, a greatly strengthened partnership between the natural sciences and the social sciences. Nowhere is the case for mutual respect and constructive collaboration between natural scientists and social scientists more persuasive than in efforts to come to terms with global change. While general circulation models are obviously important to the study of climate change, for example, the value of their results is sharply limited in the absence of information about rates of emissions of greenhouse gases into the atmosphere resulting from human actions. Similarly, forecasts relating to the impact of the depletion of stratospheric ozone on human health are sensitive to information regarding the actions people are likely to take to block the flow of ultraviolet radiation reaching the surface of the earth or to protect themselves from the harmful effects of ultraviolet radiation.

Significant barriers currently obstruct effective collaboration between natural scientists and social scientists interested in global change. In addition to problems of terminology that impede communication and attitudinal problems that operate to lower mutual regard, existing incentive structures offer few rewards for members of either community to expend time and energy on efforts to work collaboratively with members of the other. More productive collaboration between natural scientists and social scientists on issues of global change will consequently require both the initiation of research activities that compel individual scientists to interact with each other on a sustained basis and a deliberate effort to structure incentives to reward those who seek to improve research collaboration between the natural sciences and the social sciences with respect to the global change research agenda.

In addition, careful attention must be given to appropriately organizing the federal government's research on the human dimensions of global change. In our view, the activities of the Committee on Earth and Environmental Sciences (CEES) represent a major step forward in interagency coordination regarding global change research. We believe that the efforts of this committee to integrate the human dimensions into the overall U.S. Global Change Research Program (USGCRP) are not only appropriate in planning terms but also conducive to strengthening the partnership between natural scientists and social scientists working in this field. Our committee notes as well that some of the necessary research on the human dimensions of global change, particularly concerning human responses to and consequences of global change, would be appropriate for support by the newly emerging Mitigation and Adaptation Research Strategies program being proposed by CEES.

A significant barrier to an effective research effort is the current division of labor among federal agencies, which is such that almost no agency supporting research on global change has either the responsibility or the human resources to mount a significant program of social science research (NSF is a notable exception). Accordingly, mission agencies are not now in a position to manage a research effort in which strengthening the partnership between the natural sciences and the social sciences is a priority. This suggests to us that to achieve long-term success in human–environment research, action must be taken above the level of individual agencies. The Committee on Earth and Environmental Sciences might, as appropriate, assign important areas of human–environment research to NSF, or to particular mission agencies with the requirement that they take on new staff or make use of outside expertise to handle the assignment. As another alternative, it might create a new organizational entity, appropriately staffed with social and natural scientists, to conduct or manage basic and applied research on issues involving human interactions that are unlikely to be addressed in any systematic way by existing agencies.

RESEARCH PRIORITIES

Social scientists employing well-developed theories and methods have made significant contributions to our understanding of the rudiments of social driving forces leading to global change and the impacts of environmental changes on humans at the local, regional, and national levels. Further applications of existing approaches can continue to yield useful insights.

At the same time, the global change research agenda poses challenges for the understanding of human behavior and social institutions that require extraordinary efforts to push beyond existing disciplinary and theoretical categories. To broaden and deepen our understanding of the human dimensions of global change, research must transcend the boundaries of existing disciplines and research traditions. To illustrate, research dealing with individual or collective choice, or social conflict about how to respond to global change, lends itself to collaborative efforts on the part of anthropologists, economists, political scientists, psychologists, and sociologists. Researchers will need to develop concepts linking human actions to their cumulative and long-term consequences, procedures to relate local decisions to outcomes on regional and global scales, and methods to assess the interactions among changes in human population, technology, economic forces, social organization, and policy.

To realize the full potential of the research community to contribute to our understanding of global change, the committee concludes that research should proceed along two tracks simultaneously. As a result, we have separate recommendations to offer regarding programs of investigator-initiated research and programs of targeted or focused research on the human dimensions of global change.

Recommendation 1 The National Science Foundation should increase substantially its support for investigator-initiated or unsolicited research on the human dimensions of global change. This program should include a category of small grants subject to a simplified review procedure.

We applaud the recent initiative of the National Science Foundation in setting up a special competition for research proposals dealing with the human dimensions of global change. In our view, this program of investigator-initiated research should be established on a long-term basis, structured to include the full range of social and behavioral sciences, and expanded substantially in terms of funding. It should be open to scientists located at universities and other research centers. The program should accept proposals approaching global change issues in terms of well-established research traditions as well as newer and more innovative methodologies.

In the following paragraphs, we present a number of criteria of evaluation that should inform the thinking both of those preparing proposals for submission and of those charged with reviewing and making choices among proposals under the terms of this program. Needless to say, NSF should continue to apply its standard criteria for evaluating the quality of submissions. Among proposals of high quality, NSF should in addition rely on the following guidelines for selection, Of course, individual projects cannot respond to all these concerns simultaneously; the guidelines should be applied flexibly in making decisions about the relative merits of competing proposals.

1a: Studies of the anthropogenic sources of global change deserve priority to the extent that they address human actions that have a large impact on one or more of the major global environmental changes. As Chapter 3 shows, not all the proximal causes of anthropogenic change are of equal significance. For example, in the case of climate change, emissions of greenhouse gases loom large, and the burning of fossil fuels is the largest source of these emissions by a considerable margin. More specifically, certain functional areas (for example, electricity generation and motorized transport) stand out as major contributors. It follows, in our view, that projects designed to focus on such major proximal causes should receive priority over those concerned with lesser or more marginal sources of anthropogenic change in large natural systems. The committee is aware of the great scientific uncertainties that sometimes exist about the relative importance of different human activities as proximate causes of particular global changes. Human activities that seem to be of only modest importance today may loom much larger in the future, either because they have increased greatly in magnitude or because new knowledge shows that they are more important than previously thought. Although the impact criterion can only be applied within limits of uncertainty, it should still be applied as one measure of the merits of proposals for investigator-initiated research.

1b: Studies of the anthropogenic sources of global change should receive priority to the extent that they emphasize interactions among social driving forces. A sophisticated understanding of anthropogenic changes in large environmental systems requires consideration of forces operating in human systems that are typically studied by separate disciplines. Not only do individual factors interact with each other, but there are also intervening variables that play important roles in determining the ultimate impact of the primary forces. The significance of population growth as a source of anthropogenic change, for instance, depends on other factors, such as the level of economic development, the degree of urbanization, residential patterns within human settlements, and distances between settlements. It follows, in our judgment, that research on the human dimensions of global change should give priority to probing such multivariate relationships as a means of improving our grasp of the anthropogenic sources of global change.

1c: While there is a place for global-level studies, the emphasis in the near term should fall on comparative studies at the national, regional, and local levels. Whether or not there are identifiable links at the global level of aggregation between driving social forces (such as population or economic development) and anthropogenic changes (such as increases in the level of carbon dioxide in the atmosphere), there may well be strong relationships among such variables at the national, regional, and local levels. We note, as well, that a variety of causal mechanisms can give rise to similar forms of anthropogenic change and that different impacts on environmental systems can arise from apparently similar human actions. To illustrate, the destruction of tropical moist forests in Brazil, Indonesia, and Zaire have similar environmental consequences (for example, the extinction of species and the release of carbon stored in trees), but the driving social forces leading to these outcomes are by no means identical in the three cases. The patterns of energy usage in Eastern and Western Europe, by contrast, are substantially different, despite similarities of resource base, history, and culture. Under the circumstances, the way to avoid drawing incorrect conclusions about one level of analysis from observations at another level is to adopt a comparative perspective, examining similar occurrences in a variety of settings and differentiable outcomes in similar settings to probe the nature of the causal mechanisms at work. The theory, methods, and data available to social scientists are best developed at these levels of analysis. At the same time, certain human interactions with global change do occur at a global scale (e.g., the impact of the spread of scientific knowledge); they deserve special study as forms of global social change. The emphasis should be on global comparative research, that is, studies at a single level of analysis that examine the range of variation in the relevant phenomena around the world and studies tracing the relationship across time between global social change and global environmental change.

1d: Although there is room for analyses on different time scales, there is a need to be especially supportive of studies dealing with time scales of decades to centuries. Studies oriented to time scales of a decade or less are likely to focus on factors such as short-term fluctuations in public opinion, the impact of business cycles, and the introduction of particular public policies (for example, favorable tax treatment accorded to ranchers in Brazil). Studies framed in terms of time scales of decades to centuries, by contrast, will focus on different classes of events, such as the Industrial Revolution, long-term patterns of urbanization, and the gradual evolution of systems of private property rights. We believe that studies of shorter-term phenomena have an important place in improving understanding of the human dimensions of global change. Nonetheless, we are convinced that there is a compelling case for devoting more attention to the longer time scales in analyzing the human dimensions of global change. By and large, modern work in the social sciences has paid far more attention to short time scales. Yet global change is intrinsically a historical phenomenon that calls for an examination of long-term changes in human systems as well as environmental systems.

1e: There is a need to support studies that compare interventions at different points in the cycle of human–environment relationships and make empirical assessments of their relative effects. Human responses to global changes vary on several important dimensions. There are significant differences, for example, between responses to actual and anticipated changes and between deliberate and incidental responses. In addition, there are numerous points in the cycle of interactions between human systems and environmental systems at which humans can intervene to protect their values. Chapter 4 differentiates among three types of mitigation and among blocking, adjustment, and efforts to enhance the robustness of social systems. Research on the human dimensions of global change should seek both to identify the factors that determine when humans will opt for one or another of these responses and to assess the benefits, costs, and unanticipated effects of different types of responses in a variety of socioeconomic and political settings.

1f: Research should make a systematic effort to compare and contrast human responses at different levels of social organization. Responses to changes in environmental systems by individuals, corporations, communities, and countries may vary in terms of both the ease with which they are initiated and the consequences of pursuing them. Different types of responses may also be available at various levels of social organization. There is much to be learned about factors controlling levels of human response to global environmental changes and determining the extent to which humans are able to coordinate their actions across several levels of response at the same time. It follows that there is a need for studies that examine two or more levels of human response on a comparative basis.

1g: There is much to be gained from studies that differentiate among distinct methods or mechanisms for influencing human behavior. Human societies have developed a variety of mechanisms for diverting behavior away from disruptive patterns and channeling it toward productive or at least benign patterns. These mechanisms may be hierarchical, as in regulations promulgated by public authorities, or decentralized, as in the supply and demand relationships of markets. They may be traditional, as in indigenous systems for managing common property resources, or modern, as in resource regimes devised by legislatures. They differ as well in terms of the mix of rewards and punishments they use to influence human behavior. There is a need to learn more about these mechanisms as applied to the problems posed by global environmental changes and to think about the consequences they are likely to produce in terms of effectiveness, efficiency, and equity.

1h. There is a need for studies of the robustness of human systems (including social, technical, agricultural, economic, and political systems) in the face of global environmental change. Human systems are robust in the face of global change to the extent that they can adapt to perturbations in environmental systems without extreme hardship. Robustness may involve resilience, that is, a return to a previous state, or it may involve adaptations that lead to a new, but desirable, state. Examples of robustness include the ability to substitute energy conservation for fuels under conditions of shortage or one crop for another under conditions of drought. Though it is easy enough to provide examples, the concept of robustness and related concepts such as resilience and resistance are still in need of development for applications to human systems. Studies of the ways human systems change in response to environmental change can help develop those concepts and build knowledge about the determinants of robustness in complex social systems. As changes in environmental systems become more pervasive, the need to improve our understanding of robustness in social systems will increase rapidly.

1i: Proposals deserve priority to the extent that they are likely to enhance understanding of processes of decision making and

conflict management in response to global environmental changes.
Because knowledge about likely global changes is always incomplete or imperfect and because alternative responses will affect human values and interests in different ways, conflicts are inevitable at whatever level decisions are taken. Rational responses to global environmental changes will consequently require the establishment of institutional arrangements to evaluate available knowledge relating to such changes, facilitate decision making about responses to the changes, and manage the resultant social conflicts. Given the widespread frustration associated with policy making concerning environmental issues and the magnitude of the human responses needed to address many global changes, a concerted effort to improve the quality of collective decision making in this area is warranted.

1j: Special attention should be given to proposals that suggest effective methods of enhancing the partnership between the natural sciences and the social sciences or encouraging interdisciplinary research efforts among the social sciences relating to global environmental change. We discuss ways to train scientists and structure rewards to encourage interdisciplinary research in a later section of this chapter. Here we simply note that research projects differ in the extent to which they provide opportunities for overcoming the barriers to collaboration among the various sciences. A project dealing with the design of institutions for the sustainable use of renewable resources such as forests or fisheries, for example, offers more scope for collaboration between natural scientists and social scientists than a project dealing with the behavior of public officials in environmental regulatory agencies. In our judgment, preference should go to proposals that seem especially promising as vehicles for bridging these disciplinary gaps.

1k: Proposals deserve serious consideration to the extent that they include effective plans for increasing international collaboration. The case for international collaboration in this field of research rests on several considerations. Such collaboration can contribute to fulfilling the need for comparative studies noted earlier in this section. Equally important, collaboration with counterparts in other countries can expose researchers to different intellectual traditions and to the divergent conceptual, methodological, epistemological, and normative premises embedded in these traditions. Because of the global nature of the changes of interest, it is essential to overcome intellectual parochialism in dealing with this subject. In our judgment, there is much to be said for promoting active collaboration between U.S. researchers and the global change research communities in other countries that are allocating resources to this field of study.

Recommendation 2   The National Science Foundation, other appropriate federal agencies, and private funding sources should establish programs of targeted or focused research on the human dimensions of global change.

The committee concludes that there is a national need to establish ongoing programs of targeted or focused research—that is, programs that will concentrate resources to advance our understanding of topics selected by the funding sources for their obvious significance for global environmental change. It will be important to establish procedures for reviewing and updating the topics included in these programs from time to time. All topics selected for focused research should meet the following criteria:

—they deal with matters that are of first-order significance to understanding causes, consequences, and responses to global environmental change;
—they raise questions that typify larger classes of concerns relating to the human dimensions of global change;
—they address one or more of the major categories of global environmental change;
—they show promise of yielding timely advances regarding questions of broad interest to the social sciences.

At the same time, the topics chosen must be sufficiently well defined to provide a basis for targeted research.

In our judgment, these programs of targeted research on the human dimensions of global change should resemble the focused programs that the National Science Foundation currently operates in other areas (for example, atmospheric sciences, long-term ecological research, Arctic systems science). Among other things, this suggests the value of establishing scientific advisory panels to provide outside guidance for the managers of these focused programs. While scientists located at any research center should be encouraged to submit proposals to these focused programs of research, we anticipate that many proposals will come from the national centers for research on the human dimensions of global change recommended in a later section of this chapter. Proposals submitted to targeted programs should be subject to the guidelines outlined under recommendation 1. In addition, any project funded through the focused programs will necessarily deal directly with at least one of the substantive topics spelled out in the relevant program announcement. Programs of focused or targeted research might be located either in the NSF or in appropriate mission agencies. As already noted, programs in mission agencies should have the guidance of social scientists drawn from agency staff or outside advisory groups.

The committee has not conducted an exhaustive review of all potential topics for the programs of targeted research recommended here. But after extensive discussion, we agree that each of the following topics meets our criteria for inclusion in the initial phase of focused research dealing with the human dimensions of global change. Thus, the following are good examples of appropriate focused research programs. Several of them could be managed by federal mission agencies with appropriate staffing and oversight.

2a: Energy Intensity. Why do economies differ so markedly in their energy intensity? How and why does the consumption of energy per unit of gross national product change over time? What do the answers to these questions imply about opportunities to reduce carbon dioxide emissions? Carbon dioxide is the single, largest contributor to the greenhouse effect that underlies projections of global climate change over the next several decades; the use of fossil energy to drive agricultural and industrial production is the largest source of anthropogenic carbon dioxide. Not only do the economies of nations differ greatly in their energy intensity, but there has also been considerable change in the energy intensity of some, but not all, economies over the last 20 years. An understanding of the sources of these variations is important in its own right. Research on this issue can provide insights as well into other aspects of the global industrial metabolism. Since changes in energy intensity in recent years have been, at least in part, a response to shortages and to the development of new technologies, research on this topic can contribute to our general understanding of the ways in which socioeconomic systems respond to environmental changes.

2b: Land Use and Food Production. What factors change systems of land use and food production toward either degradation of resources or sustainability? How do such changes correlate with population growth, technological development, and the evolution of social institutions? Changes in patterns of land use, resulting in the conversion of forests and wetlands and in the introduction of irrigation systems, constitute one of the principal sources of climate change and the loss of biodiversity. Many of these changes are related to the production of food for human consumption. By comparing systems that vary along a continuum from extensive to intensive patterns of land use, it is possible to elucidate the social driving forces leading to change in patterns of land use. This leads directly to a consideration of sustainability treated as the maintenance of the biogeochemical condition of soils and vegetation coupled with reliable crop production for users over the long term.

2c: Valuing Consequences of Environmental Change. What alternative approaches are available for use in valuing consequences of environmental changes not well reflected in market prices? What institutional arrangements would we need to establish to ensure the effective use of the most promising of these approaches? Many of the large-scale consequences of anthropogenic changes in environmental systems are subject to a high degree of uncertainty, occur over long time periods (sometimes involving several human generations), and involve values that are not captured in a clear-cut way in market prices. Consequently, societies lack unambiguous indicators of the social costs and benefits of many global environmental changes. The long-term health effects of increased ultraviolet radiation illustrate the problems of measuring intertemporal distributions of benefits and costs. Biological diversity constitutes an example of nonmarketed resources under threat of serious degradation. Loss of geopolitical position resulting from unfavorable impacts of climate change on agricultural productivity exemplifies another class of nonmarket effects. Research on methods of valuing these kinds of consequences—especially methods capable of producing at least ordinal rankings—could yield high returns measured in terms of improving the quality of public discussion of appropriate ways of responding to global environmental changes. Valuation research should explicitly address the subjective nature of valuation and the phenomenon of differences in valuation, for instance, by exploring ways of soliciting valuations from different actors as part of the social decision process.

2d: Technology–Environment Relationship. What determines whether technologies developed and adopted in industry, agriculture, and other economic sectors mitigate or exacerbate global environmental change? What are the roles of factor prices, regulatory practices, systems of property rights, standards of performance, and other characteristics of the decision environment in determining which technological options are pursued and adopted? Basic research on conditions governing the occurrence and diffusion of innovations can provide a basis for analyzing the focused question of what institutional factors influence the extent to which successful innovations have beneficial or detrimental environmental effects. Progress in this area has great potential to underpin the development of policies aimed at mitigating environmental change or increasing the robustness of social systems in the face of change.

2e: Decision Making in Response to Global Environmental Change. How do individuals, firms, communities, and governments come to perceive changes in environmental systems as requiring action? How do they identify possible responses and assess the probable consequences of such responses? Are there cultural differences in the way human communities deal with such issues? Uncertainty is a prominent feature of most global environmental changes. Because of the complexity of large physical and biological systems, it is often impossible to predict major changes in environmental systems; it is even hard to attach probabilities with any confidence to different possible trajectories of change. To understand human responses to global change, therefore, we need to learn more about individual and collective attitudes toward risk, factors affecting propensities to launch anticipatory responses, and the complexities of collective decision making when consequences may be profound but not experienced until much later. In this context, it may prove helpful to consider the literature on risk assessment and human behavior in the face of natural hazards as sources of insights into human responses to global environmental changes.

2f: Environmental Conflict. How will global environmental changes intensify existing social conflicts or engender new forms of conflict? What techniques of conflict resolution or conflict management are likely to prove effective in coming to terms with these conflicts? Global environmental changes are likely to cause major realignments of ideologies and interests and, in the process, to intensify existing conflicts and precipitate new forms of social conflict among groups espousing divergent belief systems, holding different value priorities, or pursuing incompatible interests. There is a need to think systematically about conflicts arising from the expectation that future global change will benefit some social groups, countries, or generations at the expense of others; from direct consequences of global environmental change, such as migrations of environmental refugees or pressures to redraw borders in the face of changes in agricultural productivity; and from long-range or indirect consequences, such as disruption of ecosystems due to acid precipitation or loss of global biological diversity resulting from activities occurring within the domestic jurisdictions of single countries. Some of these types of conflict will prove resistant to resolution through ordinary procedures for handling social conflict, such as diplomacy or negotiation. It is therefore important to consider the effectiveness of alternative approaches to conflict resolution or management that may help in dealing with specific categories of conflicts arising from global environmental changes.

2g: International Cooperation. What can we learn from the recent experience with marine resources, ozone depletion, and transboundary pollution that is relevant to international efforts to deal with climate change and the loss of biodiversity? When do governments resort to international cooperation in dealing with environmental changes, and when are the resultant regimes likely to prove effective? Efforts to cooperate in coming to terms with environmental changes raise questions about collective-action problems occurring at all levels of social organization. Recently, there has been a striking growth of research interest in these problems at the international level. Partly, this reflects a widespread conviction that international cooperation will be necessary to solve all the major problems arising from global environmental changes. It also stems from a sense that the study of international cooperation is an area in which major advances in understanding of collective-action problems in general are now within our grasp.

DATA NEEDS

Like others who have wrestled with the problem, the committee has found it difficult to arrive at simple and straightforward answers to questions regarding data needs for global change research. Given the nature of the subject, it is possible to make a strong case for the relevance of a wide array of data sets in studies of global environmental change. Yet it is not feasible to collect and disseminate data on everything that may prove important for global change studies, because of the extreme cost. Nor is it easy to resolve the organizational issues arising in this field. The argument for centralization in acquiring and disseminating data relating to global change rests on grounds of standardization and efficiency. The counterarguments concern the dangers of entrusting this function to those with little understanding of particular areas, including the human dimensions of global change or, worse, to those having no direct stake in the progress of research at all. Despite the difficulties, the committee has made certain judgments regarding data needs for a national research program on the human dimensions of global change.

Recommendation 3   The federal government should establish an ongoing program to ensure that appropriate data sets for research on the human dimensions of global change are routinely acquired, properly prepared for use, and made available to scientists on simple and affordable terms.

There is a national need to (i) inventory existing data sets relevant to the human dimensions of global change, (ii) critically assess the quality of the most important of these data sets, (iii) make determinations about the quality of data required for research on major themes, (iv) investigate the cost-effectiveness of various methods of improving the quality of critical data sets, and (v) make decisions regarding new data needed to underpin a successful program of research.

There is no dearth of data relevant to the human dimensions of global change. In some areas, in fact, the quantity and quality of data available to social scientists compare favorably with the data available to natural scientists. Yet little has been done to take stock of the data currently available and, especially, to consider data problems in the light of careful judgments regarding the types and quality of data needed to conduct innovative research. It follows that we need to know more about what is already available to scientists working in this field.

The committee also recognizes that the collection and dissemination of data relating to the human dimensions of global change could become a bottomless pit in terms of time, energy, and financial resources. There is therefore a need to establish priorities and to devise economizing strategies in this area. To illustrate, comparative work using countries as the unit of analysis can tap into the vast array of statistical data on economic, political, and social factors already collected and organized for analysis at the country level. For some work of this kind, especially when new country-level data must be developed, it may make sense to adopt a strategy of always including a few countries that are large and pivotal in environmental terms (for example, the United States, the Soviet Union, two or three European countries, China, India, and Brazil) but merely sampling other countries.

While individual investigators will continue to play key roles in the collection and analysis of data, there is a persuasive case for creating a federal program to deal with the issues identified in the preceding paragraphs. Public agencies collect the bulk of the relevant data in the first place; they should be made aware of the data requirements of those working on the human dimensions of global change in planning their data collection strategies. In addition, the magnitude of the issues relating to the quantity and quality of relevant data is such that individuals or private groups cannot hope to handle them effectively or efficiently. For these reasons, we recommended in Chapter 6 that the information network for global research include the human dimensions and that the federal government seriously consider establishing a national data center on human dimensions with appropriate independent oversight.

Members of the committee expressed mixed feelings about several current proposals for mechanisms to collect and disseminate data on all aspects of global change and, more particularly, about NASA's Earth Observing System (EOS) program and its Data and Information System (EOSDIS). There is widespread agreement on the need to make data and information relating to global change easily accessible to all interested users on an inexpensive basis. But opinions vary regarding the desirability of creating a national "information utility" in this area. In part this reaction reflects differing assessments of actual or potential deficiencies in existing mechanisms for collecting and disseminating data. In part it stems from deeper differences about the relative merits of centralized arrangements organized and operated by government agencies in contrast to decentralized arrangements provided by private operators in response to rising demand. The committee notes as well that a large fraction of the total U.S. Global Change Research Program focused budget is slated to go to NASA and that the EOS program will consume most of this funding, facts that raise questions about the achievement of a proper balance in allocating funds among global change research priorities.

As noted in Chapter 6, the committee recommends that social scientists should be included in all phases of the design and operation of EOS and EOSDIS. Such an approach is critical to strengthening the partnership between the natural sciences and the social sciences as recommended earlier in this chapter. What is more, the study of human dimensions has important needs that should be taken into account in developing EOS and EOSDIS. Appropriately collected remotely sensed data, for example, can be used creatively to address the human dimensions, and this potential can be expanded with efforts to validate remote indicators of human activity against ground data. In addition, many of the data requirements for research on the human dimensions of global change involve improved methods for ground observation. For this reason, we recommend that the social sciences be strongly represented on science advisory panels for EOS and EOSDIS.

HUMAN RESOURCES AND ORGANIZATION

No amount of planning for the conduct of research on the human dimensions of global change can produce significant results in the absence of a community of social scientists ready, able, and willing to make a commitment to this field of study. In part this is a matter of motivating or capturing the interest of individual scientists. In part, however, it has to do with the actions of organizations in rewarding or discouraging those desiring to work on global change issues and in providing the support needed for designing and implementing research projects in this area. In this section, we address two major program elements dealing with the interlocking issues of human resources and organization.

Recommendation 4   The federal government, together with private funding sources, should establish a national fellowship program. Through it, social and natural scientists prepared to make a long-term commitment to the study of the human dimensions of global environmental change could spend up to two years interacting intensively with scientists from other disciplines, especially scientists from across the social science–natural science divide.

In the university world, academic departments organized around established disciplines control the rewards available to researchers. Those who wish to succeed in this world consequently experience strong incentives to develop courses that fit into the mainstream of their home disciplines and to conduct research whose products are publishable in the most prestigious journals of these disciplines. While this institutional structure should not be thought of as completely impervious to change, we do not expect any fundamental transformation of the existing order during the foreseeable future. Over the short term, at least, researchers interested in the human dimensions of global environmental change will find themselves operating within these structural constraints.

It is imperative to find ways to allow individual scientists to push beyond the boundaries of their home disciplines in thinking about global change without jeopardizing their career trajectories. One way to accomplish this goal is to create a prestigious nationwide fellowship program designed to allow individuals desiring to retool or enhance their knowledge of the human dimensions of global change to spend considerable periods of time (up to two years) interacting intensively with social scientists and natural scientists from other disciplines working in the area. Such fellowships should be open to advanced graduate students, postdoctoral scientists, and mid-career scientists on a competitive basis. They should carry large enough stipends to attract the best and the brightest and be prestigious enough to count heavily in evaluations of career performance. While individual fellows should be allowed considerable freedom to design their own programs, we expect that many will choose to associate themselves with one of the national centers for research on the human dimensions of global change called for in our next recommendation.

Recommendation 5   The federal government should join with private funding sources to establish about five national centers for research on the human dimensions of global change and to make a commitment to funding these centers on a long-term basis.

There is no surefire method of getting results in any field of scientific inquiry. Every method has its drawbacks; no method by itself is sufficient to do the job. Nonetheless, the committee believes that, in the social sciences at least, well-funded national research centers have played an important role in producing some of the most impressive advances in both basic and applied research. Exemplary cases include the Office of Population Research at Princeton in the field of demography, the National Bureau of Economic Research in quantitative macroeconomics, and the complex of research centers at Harvard and the Massachusetts Institute of Technology in the area of arms control. Not only have such centers provided critical mass by bringing together groups of individual scientists with overlapping interests, but they have also served to surmount some of the institutional impediments to progress in interdisciplinary or transdisciplinary fields of study. *The human dimensions of global change is an emerging field of inquiry that is ripe for this sort of treatment; about five national centers for research in this general area should be established over the next 3-5 years.*

Those involved in the establishment of these centers should be guided by several important considerations. It is essential to ensure that the centers are able to survive on a long-term basis so that they can concentrate on the subject matter at hand rather than being forced to fight for their institutional survival on a continual basis. A smaller number of centers with institutional and financial security would be preferable, we believe, to a larger number plagued with insecurity. Every effort should be made to organize the centers in such a way as to strengthen the partnership between social scientists and natural scientists working on global change issues. Whether this implies appointing natural scientists to the staff of each center, collocating the centers with strong programs on the role of environmental systems in global change, or some combination of these approaches is a matter best left to those responsible for drawing up detailed blueprints for individual centers. The goal of strengthening the links between the natural sciences and the social sciences is a central rationale for having research centers.

There are, in addition, questions relating to the division of labor among centers. The substantive centers should not seek to duplicate the work of the national data center discussed in the preceding section. Rather, the national data center should interact with each of the substantive centers to ensure that relevant data are collected and made available in an appropriate manner. The question of a division of labor among the substantive centers themselves is another matter. It seems doubtful to us that the basic research needed in this area can be clearly divided among the centers. Yet it may make sense to direct the attention of individual centers toward particular topics at the applied level. For instance, some centers may devote particular attention to global environmental changes that are systemic in nature, like climate change or ozone depletion, while others deal more with cumulative changes, like the loss of biodiversity or desertification.

We have identified several distinct models that are worthy of consideration by those responsible for setting up the national centers: (i) interdisciplinary research centers located at individual universities (for example, the Office of Population Research at Princeton), (ii) semiautonomous centers loosely affiliated with individual universities (for example, the Scripps Oceanographic Institution), (iii) physically separate centers operated by consortiums of universities (for example, the National Center for Atmospheric Research), (iv) completely independent or freestanding centers (for example, Resources for the Future or the Rand Corporation), and (v) government operated research centers managing both intramural and extramural research programs (for example, the National Institutes of Health).

In our judgment, there is a persuasive case for maintaining relatively close ties between universities and the national centers. Most of the theoretically significant advances in our understanding of human behavior have occurred in university settings in contrast to independent centers, which show a marked tendency to focus more on applied research and to become associated with one or another ideological perspective. At the same time, there may well be topics relating to the human dimensions of global change that lie at the intersection between basic and applied research and that are therefore most appropriately investigated at a government-operated research center, like the proposed National Institutes of the Environment.

FUNDING

What would this national research program on the human dimensions of global change cost? While the committee did not regard the drawing up of detailed cost estimates as part of its mandate, we can provide a good sense of the order of magnitude of funding needed to implement the program we have recommended. In arriving at the estimates reported in the following paragraphs, we have been guided by the following assumptions. First, it makes sense to phase in the program over a period of several years; the figures we provide for a program in full-swing should not be interpreted as a discontinuous jump from one fiscal year to the next. If the funding is available, the phase-in could be accomplished in three years. Second, we assume that all five program elements are necessary to forge a comprehensive national research program on the human dimensions of global change and that it is important to strike a proper balance among the individual program elements with regard to funding. Third, we have turned for guidance, wherever appropriate, to the focused budget of the U.S. Global Change Research Program for fiscal 1991.

Investigator-initiated research on human interactions was funded in 1991 at a level of $3.6 million per year (through NSF). We believe that this program element can and should be tripled to a level of about $11 million. It is difficult to make an accurate estimate of the current level of funding for targeted or focused research on the human dimensions of global change. In our judgment, however, there is a convincing case for funding research of this sort at a level comparable to that for investigator-initiated research. This would add about another $11 million per year at the end of the phase-in period.

A fellowship program in full operation that awarded 100 two-year fellowships per year to graduate and postdoctoral students and midcareer scholars would cost $10 million per year if the average annual cost were $50,000 per fellowship, including indirect costs. A commitment to provide each of the national centers with a small but strong core staff and a measure of stability within their larger institutional settings could be maintained for about $1 million per center per year. Five such centers could be funded for $5 million a year.

The matter of funding for data acquisition and dissemination is, in many ways, the hardest to resolve. There are good reasons to believe both that this program element can absorb almost unlimited funds and that there is a tendency to favor the allocation of resources to this function as the lowest common denominator among decision makers who disagree on substantive priorities. Nonetheless, we believe that the fiscal 1991 budget, in which a little over 20 percent of the funds allocated to human interactions research goes to data acquisition and dissemination, reflects an appropriate judgment concerning the priority to be given to data issues in a national research program on the human dimensions of global change. On this basis, we recommend that funding for the acquisition and dissemination of data on the human dimensions be increased over the transition period to a level of $8-10 million.

Adding up these estimates for the individual program elements brings us to our final recommendation:

Recommendation 6   The federal government should increase funding for research on the human dimensions of global change over a period of several years to a level of $45-50 million.

The committee has concluded not only that this level of support would make possible the establishment of a balanced national research program on the human dimensions of global change but also that the research community will be able to take on such a commitment over a three-year period if the funding is available. For the sake of comparison, this level of funding would represent about 5 percent of the fiscal 1991 budget for the U.S. Global Change Research Program (USGCRP) or 4 percent of the proposed fiscal 1992 budget, in contrast to the 3 percent currently budgeted. In light of the National Research Council's conclusion that the human interactions science priority is "the most critically underfunded in the fiscal 1991 budget for the USGCRP" (National Research Council, 1990c:95), an increase of this magnitude over a short time period seems fully justified. In our view, support for appropriate parts of the research program outlined here could come from an emerging Mitigation and Adaptation Research Strategies program as well as from the Global Change Research Program.

References

Agarwal, A.
 1990 The North-South perspective: Alienation or interdependence. *AMBIO* 19(2):94–96.
Aitken, S.C., et al.
 1989 Environmental perception and behavioral geography. In Gary L. Gaile and Cort J. Wilmott, eds., *Geography in America*. Columbus: Merrill.
Allen, J.C., and D.F. Barnes
 1985 The causes of deforestation in developing countries. *Annals of the Association of American Geographers* 75:163–184.
AMBIO
 1989 Importance of space observations for global change. *AMBIO* 18(4):25–35.
Anderson, A.B.
 1990 Deforestation in Amazonia: Dynamics, causes, and alternatives. Pp. 3–23. in A.B. Anderson, ed., *Alternatives to Deforestation: Steps Toward Sustainable Use of the Amazon Rain Forest*. New York: Columbia University Press.
Anderson, Jr., O.E.
 1953 *Refrigeration in America: A History of a New Technology and Its Impact*. Princeton, N.J.: Princeton University Press.
ARCS Workshop Steering Committee
 1990 Arctic Systems Science: Ocean-Atmosphere-Ice Interaction. Washington, DC: Joint Oceanographic Institutes, Inc.
Ascher, W.
 1978 *Forecasting: An Appraisal for Policy-Makers and Planners*. Baltimore, Md.: Johns Hopkins.

Aubert, V.
1963   Competition and dissensus: Two types of conflict and conflict resolution. *Journal of Conflict Resolution* 7:26–42.

Ausubel, J.H.
1989   Regularities in technological development: An environmental view. Pp. 70–91 in J.H. Ausubel and H.E. Sladovich, eds., *Technology and Environment*. Washington, D.C.: National Adacemy Press.

Ausubel, J.H., R.A. Frosch, and R. Herman
1989   Technology and environment: An overview. Pp. 1–20 in J.H. Ausubel and H.E. Sladovich, eds., *Technology and Environment*. Washington, D.C.: National Academy Press.

Ausubel, J.H., and H.E. Sladovich, eds.
1989   *Technology and Environment*. Washington, D.C.: National Academy Press.

Axelrod, R.
1984   *The Evolution of Cooperation*. New York: Basic Books.

Axline, W.A.
1978   Integration and development in the commonwealth Caribbean: The politics of regional negotiations. *International Organization* 32:953–973.

Ayres, R.U.
1978   *Resources, Environment, and Economics: Applications of the Materials/Energy Balance Principle*. New York: Wiley.

Ayres, R.U., and S.R. Rod
1986   Reconstructing an environmental history: Patterns of pollution in the Hudson-Raritan basin. *Environment* 28(4):14–20, 39–43.

Bacard, A.
1989   The second genesis: Future technologies and humanism. *The Humanist* 49:9–11.

Baksh, M.
n.d.   Changes in Machiguenga quality of life. Manuscript, University of California, Los Angeles.

Balée, W. and Gély, A.
1989   Managed forest succession in Amazonia: The Ka'apor case. *Advances in Economic Botany* 7:129–158.

Bardach, E., and R.A. Kagan
1982   *Going by the Book*. Philadelphia: Temple University Press.

Barnett, H.J., and C. Morse
1963   *Scarcity and Growth: The Economics of Natural Resource Availability*. Baltimore, Md.: The Johns Hopkins Press for Resources for the Future.

Barton, A.H.
1969   *Communities in Disaster*. Garden City, N.Y.: Anchor-Doubleday.

Bates, R.
    1980   *States and Markets in Africa.* Berkeley, Calif.: University of California Press.
Baum, A., and P. Paulus
    1987   Crowding. In D. Stokols and I. Altman, eds., *Handbook of Environmental Psychology.* 2 volumes. New York: Wiley.
Baumgartner, T., and A. Mudttun
    1987   *The Politics of Energy Forecasting.* London: Oxford.
Baumol, W.J., and W.E. Oates
    1988   *The Theory of Environmental Policy.* New York: Cambridge University Press.
Beauchamp, T.L., and J.F. Childress
    1989   *Principles of Biomedical Ethics,* 3rd ed. New York: Oxford
Belsley, D.A., and E. Kuh, eds.
    1986   *Model Reliability.* Cambridge, Mass.: MIT Press.
Benedick, R.E.
    1989a   The ozone protocol: A new global diplomacy. *Conservation Foundation Letter* 4.
    1989b   Ozone diplomacy. *Issues in Science and Technology* 6(1)Fall:43–50.
    1991   *Ozone Diplomacy.* Cambridge, Mass.: Harvard University Press.
Bentkover, J.D., V.T. Covello, and J. Mumpower, eds.
    1985   *Benefits Assessment: The State of the Art.* Dordrecht, The Netherlands: Reidel.
Berk, R.A., T.F. Cooley, C.J. LaCivita, S. Parker, K. Sredl, and M. Brewer
    1980   Reducing consumption in periods of acute scarcity: The case of water. *Social Science Research* 9:99–120.
Berry, B.J.L.
    1991   Urbanization. Pp. 103–119 in B.L. Turner II et al., eds., *The Earth as Transformed by Human Action.* New York: Cambridge University Press.
Berry, L.
    1990   *The Market Penetration of Energy-Efficiency Programs.* ORNL/CON-299. Oak Ridge, Tenn.: Oak Ridge National Laboratory.
Binswanger, H.
    1989   Brazilian Policies that Encourage Deforestation in the Amazon. Environment Department Working Paper No. 16. Washington, D.C.: World Bank.
Black, J.S., P.C. Stern, and J.T. Elworth
    1985   Personal and contextual influences on household energy adaptations. *Journal of Applied Psychology* 70:3–21.
Blaikie, P., and H.C. Brookfield
    1987   *Land Degradation and Society.* London: Methuen.
Bollen, K.A.
    1980   Issues in the comparative measurement of political democracy. *American Sociological Review* 45:370–390.

1989   *Structural Equations With Latent Variables.* New York: John Wiley.

Borden, R.J., and J.L. Francis
1978   Who cares about ecology? Personality and sex differences in environmental concern. *Journal of Personality* 46:190–203.

Borden, R.J., J. Jacobs, and G.L. Young, eds.
1988   *Human Ecology: Research and Applications.* College Park, Md.: Society of Human Ecology.

Boserup, E.
1965   *The Condition of Agricultural Growth.* Chicago, Ill.: Aldine.
1981   *Population and Technological Change: A Study of Long-Term Trends.* Chicago, Ill.: University of Chicago Press.
1990   *Economic and Demographic Relationships in Development.* Baltimore, Md.: Johns Hopkins University Press.

Boulding, K.E.
1971   Environment and economics. Pp. 359–367 in W.W. Murdoch, ed., *Environment: Resources, Pollution and Society.* Stamford, Conn.: Sinaur.
1974   What went wrong, if anything, since Copernicus? *Science and Public Affairs* (January):17–23.

Boyd, R., and P.J. Richerson
1985   *Culture and the Evolutionary Process.* Chicago, Ill.: University of Chicago Press.

Braudel, F.
1983   *The Wheels of Commerce. Volume II. Civilization and Capitalism 15th–18th Century.* New York: Harper and Row.
1984   *The Perspective of the World. Civilization and Capitalism, Volume III.* New York: Harper and Row.
1985   *The Structures of Everyday Life. Volume I. Civilization and Capitalism 15th–18th Century.* New York: Harper and Row.

Brewer, G.
1983   Some costs and consequences of large-scale social systems modeling. *Behavioral Science* 28:166–185.

Brewer, G.D.
1986   Methods for synthesis: Policy exercises. Pp. 455–473 in W.C. Clark and R.E. Munn, eds., *Sustainable Development of the Biosphere.* New York: Cambridge University Press.

Brickman, R., S. Jasanoff, and T. Ilgen
1985   *Controlling Chemicals: The Politics of Regulation in Europe and the United States.* Ithaca, N.Y.: Cornell University Press.

Brookfield, H.C., F.J. Lian, K.S. Low, and L. Potter
1991   Borneo and the Malay Peninsula. Pp. 495–512 in B.L. Turner et al., eds., *The Earth as Transformed by Human Action.* New York: Cambridge University Press.

Brooks, H.
    1986   The typology of surprises in technology, institutions and development. Pp. 455–473 in W.C. Clark and R.E. Munn, eds., *Sustainable Development of the Biosphere*. New York: Cambridge University Press.

Browder, J.O., ed.
    1988   Public policy and deforestation in the Brazilian Amazon. Pp. 247–297 in R. Repetto and M. Gillis, eds., *Public Policies and the Misuse of Forest Resources*. New York: Cambridge University Press.
    1989   *Fragile Lands of Latin America: Strategies of Sustainable Development*. Boulder, Colo.: Westview.

Brown, L.A.
    1981   *Innovation Diffusion: A New Perspective*. New York: Sage.
    1982   An untraditional view of national security. In J.F. Reichart and S.R. Sturn, eds., *American Defense Policy*. Baltimore, Md.: John Hopkins University Press

Burton, I., R.W. Kates, and G.F. White
    1978   *The Environment as Hazard*. New York: Oxford University Press.

Burton, J.
    1978   Externalities, property rights and public policy: Private property rights or the spoilation of nature. Epilogue to S.N.S. Cheung, *The Myth of Social Cost*, pp. 69–91. Lansing, Mich.: The Institute of Economic Affairs.

Burton, J.W.
    1986   The history of international conflict resolution. In E.E. Azar and J.W. Burton, eds., *International Conflict Resolution: Theory and Practice*. Boulder, Colo.: Lynne Rienner.

Bush, G., and M. Gorbachev
    1990   Global environmental policy. *Environment* 32(3):12–15, 32–35.

Buttel, F.H.
    1987   New directions in environmental sociology. *Annual Review of Sociology* 13:465–488.

Butzer, K.
    1976   *Early Hydraulic Civilization in Egypt: A Study in Cultural Ecology*. Chicago, Ill.: University of Chicago Press.
    1989   Cultural ecology. In G.L. Gaile and C.J. Wilmott, eds., *Geography in America*. Columbus, Ohio: Merrill.

Campbell, G.
    1972   *Brazil Struggles for Development*. London: Charles Knight.

Campbell, J.C.
    1976   *Successful Negotiation: Trieste 1954*. Princeton, N.J.: Princeton University Press.

Carroll, J.E.
    1983   *Environmental Diplomacy*. Ann Arbor: University of Michigan Press.

Catton, W.R., Jr.
1980   *Overshoot: The Ecological Basis of Revolutionary Change.*
        Urbana, Ill.: University of Illinois Press.
Catton, W.R., and R.E. Dunlap, Jr.
1980   A new ecological paradigm for post-exuberant society. *American Behavioral Scientist* 24:15–47.
Chen, R.S., E. Boulding, and S.H. Schneider, eds.
1983   *Social Science Research and Climate Change: An Interdisciplinary Appraisal.* Dordrecht, The Netherlands: Reidel.
Chernela, J.M.
n.d.   Sustainable Development and Sustainable Control: Political Strategies of Indian Organizations in a Proposed Binational Biosphere Reserve in Ecuador and Colombia. Manuscript, Florida International University.
Chisholm, M.
1980   The wealth of nations. *Transactions of the Institute of British Geographers* 5:255–276.
1982   *Modern World Development: A Geographic Perspective.* Savage, Md.: Barnes and Noble Imports.
Ciriacy-Wantrup, S.V., and R.C. Bishop
1975   "Common property" as a concept in natural resources policy. *Natural Resources Journal* 15:713–27.
Clark, W.C.
1987   Scale relationships in the interactions of climate, ecosystems, and societies. Pp. 337–378, 474–490 in K.C. Land and S.H. Schneider, eds., *Forecasting in the Social and Natural Sciences.* Dordrecht, The Netherlands: Reidel.
1988   The human dimensions of global environmental change. Pp. 134–200 in Committee on Global Change, *Toward an Understanding of Global Change: Initial Priorities for U.S. Contributions to the International Geosphere-Biosphere Programme.* Washington, D.C.: National Academy Press.
Clarke, L.
1988   Explaining choices among technological risks. *Social Problems* 35:22–35.
1989   *Acceptable Risk? Making Decisions in a Toxic Environment.* Berkeley, Calif.: University of California Press.
Clemen, R.A.
1923   *The American Livestock and Meat Industry.* New York: Ronald Press.
Cleveland, H.
1990   The age of spreading knowledge. *The Futurist* 24:35–39.
Coale, A.J.
1970   Man and his environment. *Science* 170:132–136.
Coase, R.H.
1960   The problem of social cost. *Journal of Law and Economics* 3:1–44.

Cohen, J.L.
1985   Strategy or identity: New theoretical paradigms and contemporary social movements. *Social Research* 52:663–716.
Commoner, B.
1970   *Science and Survival.* New York: Ballantine Books.
1972   *The Closing Circle: Man, Nature and Technology.* New York: Knopf.
1977   *The Poverty of Power: Energy and the Economic Crisis.* New York: Bantam Books.
Conklin, H.C.
1954   An ethnoecological approach to shifting cultivation. *Transactions of the New York Academy of Sciences*, Series 2, 17(2):133–142.
Conway, G.R., and E.B. Barbier
1990   *After the Green Revolution: Sustainable Agriculture for Development.* London: Earthscan.
Cook, T.D., and D.T. Campbell
1979   *Quasi-experimentation: Designs and Analysis Issues for Social Research in Field Settings.* Boston, Mass.: Houghton-Mifflin.
Covello, V., and R.S. Frey
1989   Technology-based environmental health risks in developing nations. *Technological Forecasting and Social Change* 37(21 April):159.
Covello, V.T., P. Slovic, and von Winterfeldt
1987   *Risk Communication: A Review of the Literature.* Washington, D.C.: National Science Foundation.
Cox, R.W.
1983   Gramsci, hegemony, and international relations: An essay in method. *Millenium* 12:162–175.
1987   *Production, Power, and World Order: Social Forces in the Making of History.* New York: Columbia University Press.
Craik, K.H., and N.R. Feimer
1987   Environmental assessment. In D. Stokols and I. Altman, eds., *Handbook of Environmental Psychology.* 2 volumes. New York: Wiley.
Cronon, W.
1983   *Changes in the Land: Indians, Colonists, and the Ecology of New England.* New York: Hill and Wang.
1991   *Nature's Metropolis: Chicago and the Great West.* New York: W.W. Norton.
Cross, J.G., and M.J. Guyer
1980   *Social Traps.* Ann Arbor: University of Michigan Press.
Cummings, R.O.
1949   *The American Ice Harvests: A Historical Study in Technology, 1800–1918.*   Berkeley: University of California Press.

Daly, H.
  1986   On sustainable development in national accounts. In D. Collard, D. Pearce, and D. Ulph, eds., *Economics and Sustainable Environments: Essays in Honor of Richard Lecomber.* New York: Macmillan.
Daly, H.E., ed.
  1977   *Steady-State Economics.* San Francisco: W.H. Freeman.
Darley, J.M., and J.R. Beniger
  1981   Diffusion of energy-conserving innovations. *Journal of Social Issues* 37(2):150–171.
Dasgupta, P.S., and G.M. Heal
  1979   *Economic Theory and Exhaustible Resources.* New York: James Nisbet and Cambridge University Press.
Davies, J.C., V.T. Covello, and F.W. Allen, eds.
  1987   *Risk Communication.* Washington, D.C.: The Conservation Foundation.
Dawes, R.M.
  1980   Social dilemmas. *Annual Review of Psychology* 31:169–193.
Deevey, E.S., Jr.
  1960   The human population. *Scientific American* 203(3):194–204.
Denevan, W.M.
  1981   Swiddens and cattle versus forest: The imminent demise of the Amazon rain forest reexamined. *Studies in Third World Societies* 13:25–44.
Dennis, M.L., E.J. Soderstrom, W.S. Kocinski, Jr., and B. Cavanaugh
  1990   Effective disseminations of energy-related information: Applying social psychology and evaluation research. *American Psychologist* 45:1109–1117.
Dietz, T.
  1987   Theory and method in social impact assessment. *Sociological Inquiry* 57:54–69.
  1988   Social impact assessment as applied human ecology. Pp. 220–227 in R. Borden, J. Jacobs, and G. Young, eds., *Human Ecology: Research and Applications.* College Park, Md.: Society for Human Ecology.
Dietz, T., and T.R. Burns
  1991   Human agency in evolutionary theory. In B. Wittrock, ed., *Agency in Social Theory.* London: Sage.
Dietz, T., T.R. Burns, and F.H. Buttel
  1990   Evolutionary theory in sociology: An examination of current thinking. *Sociological Forum* 5:155–171.
Dietz, T., and C.M. Dunning
  1983   Demographic change assessment. In K. Finsterbusch, L.G. Llewellyn, and C.P. Wolf, eds., *Social Impact Assessment Methods.* Beverly Hills, Calif.: Sage.
Dietz, T., R.S. Frey, and E.A. Rosa
  1991   Risk, technology, and society. In R.E. Dunlap and W. Michelson,

eds., *Handbook of Environmental Sociology*. Greenwich, Conn.: Greenwood.

Dietz, T., and R.W. Rycroft
1987   *The Risk Professionals*. New York: Russell Sage Foundation.

Dietz, T., P.C. Stern, and R.W. Rycroft
1989   Definitions of conflict and the legitimation of resources: The case of environmental risk. *Sociological Forum* 4:47–70.

Dietz, T., and E.L. Vine
1982   Energy impacts of a municipal energy policy. *Energy* 7:755–758.

DiMento, J.F.
1989   Can social science explain organizational noncompliance with environmental law? *Journal of Social Issues* 45(1):109–132.

Doble, J., A. Richardson, and A. Danks
1990   *Global Warming Caused by the Greenhouse Effect*. Vol 3. in *Science and the Public: A Report in Three Volumes*. New York: Public Agenda Foundation.

Downing, T.E., and R.W. Kates
1982   The international response to the threat of chlorofluorocarbon to atmospheric ozone. *American Economic Review* 72(2):267–272.

Druckman, D.
1990   The social psychology of arms control and reciprocation. *Political Psychology* 11:553–581.

Druckman, D., B.J. Broome, and S.H. Korper
1988   Value differences and conflict resolution: Facilitation or delinking? *Journal of Conflict Resolution* 32:489–510.

Druckman, D., R. Rozelle, and K. Zechmeister
1977   Conflict of interest and value dissensus: Two perspectives. In D. Druckman, ed., *Negotiations: Social-Psychological Perspectives*. Beverly Hills, Calif.: Sage.

Druckman, D., and K. Zechmeister
1973   Conflict of interest and value dissensus: Propositions in the sociology of conflict. *Human Relations* 26:449–466.

Dunlap, R.E., and R.E. Jones
1991   Public opinion and attitudes toward environmental issues. In R.E. Dunlap and W. Michelson, eds., *Handbook of Environmental Sociology*. Greenwich, Conn.: Greenwood.

Dunlap, R.E., and W. Michelson, eds.
1991   *Handbook of Environmental Sociology*. Greenwich, Conn.: Greenwood.

Dunlap, R.E., and K.D. Van Liere
1977   Land ethic or golden rule. *Journal of Social Issues* 33(3):200–207.
1978   The "new environmental paradigm": A proposed measuring instrument and preliminary results. *Journal of Environmental Education* 9:10–19.

1984   Commitment to the dominant social paradigm and concern for environmental quality. *Social Science Quarterly* 64:1013–1028.

Dynes, R.R.
1970   *Organized Behavior in Disaster.* Lexington, Mass.: Heath.
1972   *A Perspective on Disaster Planning.* Columbus, Ohio: Disaster Research Center, Ohio State University.

Dyson, F.J.
1979   *Disturbing the Universe.* New York: Harper & Row.

Eder, K.
1985   The 'new social movements': Moral crusades, political pressure groups or social movements. *Social Research* 52:869–890.

Edney, J.J.
1980   The commons problem: Alternative perspectives. *American Psychologist* 35:131–150.

Ehrenfeld, D.
1978   *The Arrogance of Humanism.* New York: Oxford.

Ehrlich, P.R.
1968   *The Population Bomb.* New York: Ballantine.

Ehrlich, P.R., and A.H. Ehrlich
1990   *The Population Explosion.* New York: Simon and Schuster.

Ehrlich, P.R., A.H. Ehrlich, and J.P. Holdren
1977   *Ecoscience: Population, Resources, Environment.* San Francisco: W.H. Freeman and Co.

Ehrlich, P.R., and J.P. Holdren
1971   Impact of population growth. *Science* 171:1212–1217.
1972   Reviews: Dispute: *The Closing Circle. Environment* 14(3):23–26, 31–52.

Ehrlich, P.R., and J.P. Holdren, eds.
1988   *The Cassandra Conference: Resources and the Human Predicament.* College Station, Tex.: Texas A&M University Press.

Ellen, R.
1982   *Environment, Subsistence, and System: The Ecology of Small-scale Social Formations.* New York: Cambridge University Press.

Elton, C.
1958   *The Ecology of Invasion by Plants and Animals.* London: Methuen.

Emel, J., and R. Peet
1989   Resource management and natural hazards. Pp. 49–76 in R. Peet and N. Thrift, eds., *New Models in Geography: The Political-Economy Perspective.* London: Unwin Hyman.

Erikson, K.T.
1978   *Everything in Its Path.* New York: Simon and Schuster.

Erwin, T.L.
1982   Tropical forests: Their richness in coleoptera and other arthropod species. *Coleoptera Bulletin* 36(1):74–75.

1988   The tropical forest canopy: The heart of biotic diversity. Pp. 123–129 in E.O. Wilson, ed., *Biodiversity*. Washington, D.C.: National Academy Press.

Ester, P.A., and R.A. Winett
1982   Toward more effective antecedent strategies for environmental programs. *Journal of Environmental Systems* 11:201–221.

Evans, G.W., and S. Cohen
1987   Environmental stress. In D. Stokols and I. Altman, eds., *Handbook of Environmental Psychology*. 2 volumes. New York: Wiley.

Farman, J.C., B.G. Gardiner, and J.D. Shanklin
1985   Large losses of total ozone in Antarctica reveal seasonal ClO_x interaction. *Nature* 315:207–210.

Fearnside, P.M.
1989   Deforestation in Brazilian Amazonia: The rates and causes. *The Ecologist* 19:214–218.

Field, D.R., and W.R. Burch, Jr.
1988   *Rural Sociology and the Environment*. Westport, Conn.: Greenwood.

Finsterbusch, K., L. Llewellyn, and C. Wolf, eds.
1983   *Social Impact Assessment Methods*. Beverly Hills, Calif.: Sage.

Fischhoff, B.
1979   Behavioral aspects of cost-benefit analysis. In G. Goodman and W. Rowe, eds., *Energy Risk Management*. London: Academic Press.
1989   Risk: A guide to controversy. Pp. 211–319 in National Research Council, *Improving Risk Communication*. Washington, D.C.: National Academy Press.
1991   Eliciting values: Is there anything in there? *American Psychologist* 46:835–847.

Fischhoff, B.F., and L. Furby
1983   Psychological dimensions of climatics change. Pp. 180–203 in R.S. Chen, E. Boulding, and S.H. Schneider, eds., *Social Science Research and Climate Change: An Interdisciplinary Appraisal*. Dordrecht, The Netherlands: Reidel.

Fischhoff, B., B. Goitein, and Z. Shapira
1982   The experienced utility of expected utility approaches. In N. Feather, ed., *Expectations and Actions: Expectancy-Value Models in Psychology*. Hillsdale, N.J.: Erlbaum.

Fischhoff, B., S. Lichtenstein, P. Slovic, S.L. Derby, and R.L. Keeney
1981   *Acceptable Risk*. New York: Cambridge University Press.

Fischhoff, B., P. Slovic, and S. Lichtenstein
1977   Knowing with certainty: The appropriateness of extreme confidence. *Journal of Experimental Psychology: Human Perception and Performance* 3:552–564.

1981   Lay foibles and expert fables in judgments about risk. In T. O'Riordan and R.K. Turner, eds., *Progress in Resource Management and Environmental Planning.* New York: Wiley.

Fischhoff, B., O. Sverson, and P. Slovic
1987   Active responses to environmental hazards: Perception and decision making. In D. Stokols and I. Altman, eds., *Handbook of Environmental Psychology.* 2 volumes. New York: Wiley.

Fisher, R.
1964   Fractionating conflict. In R. Fisher, ed., *International Conflict and Behavioral Science: The Craigville Papers.* New York: Basic Books.

Fisher, J.L., and N. Potter
1971   The effects of population growth on resource adequacy and quality. Pp. 222–244 in National Academy of Sciences, ed., *Rapid Population Growth: Consequences and Policy Implications.* Baltimore, Md.: Johns Hopkins University Press.

Folland, C.K., T.R. Karl, and K.Ya. Vinnikov
1990   Observed climate variations and change. In J.T. Houghton, G.J. Jenkins, and J.J. Ephraums, eds., *Climate Change: The IPCC Scientific Assessment.* New York: Cambridge University Press.

Fowler, B.B.
1952   *Men, Meat, and Miracles.* New York: Julian Messner.

Fox, D.R.
1985   Psychology, ideology, utopia, and the commons. *American Psychologist* 40:48–58.

Frank, A.G.
1967   *Capitalism and Underdevelopment in Latin America.* New York: Monthly Review Press.

Freedman, D., T. Rothenberg, and R. Sutch
1983   On energy policy models. *Journal of Business and Economic Statistics* 1:24–32.

Freeman, D.M., and R.S. Frey
1986   Method for assessing the social impacts of natural resources policies. *Journal of Environmental Management* 23(3 October):229.

Freudenburg, W.R.
1984   Boomtown's youth: The differential impacts of rapid community growth on adolescents and adults. *American Sociological Review* 40:697–705.

Freudenburg, W., and E. Rosa, eds.
1984   *Public Reaction to Nuclear Power: Are There Critical Masses?* Boulder, Colo.: Westview.

Friedlander, S.K.
1989   Environmental issues: Implications for engineering design and education. Pp. 167–181 in J.H. Ausubel and H. Sladovich,

eds., *Technology and Environment*. Washington, D.C.: National Academy Press.

Fulkerson, W., D.B. Reister, A.M. Perry, A.T. Crane, D.E. Kash, and S.I. Auerbach
    1989   Global warming: An energy technology R&D challenge. *Science* 246:868–869.

Gamez, R., and A. Ugalde
    1988   Costa Rica's national park system and the preservation of biological diversity: Linking conservation with socioeconomic development. Pp. 131–142 in F. Almeda and C.M. Pringle, eds., *Tropical Rain Forests: Diversity and Conservation*. San Francisco: California Academy of Sciences.

Giedion, S.
    1948   *Mechanization Takes Command: A Contribution to Anonymous History*. New York: Oxford University Press.

Glenn, E.S., R.H. Johnson, P.R. Kimmel, and B. Wedge
    1970   A cognitive interaction model to analyze culture conflict in international relations. *Journal of Conflict Resolution* 14:35–48.

Goldstein, J.S.
    1988   *Long Cycles: Prosperity and War in the Modern Age*. New Haven, Conn.: Yale University Press.

Golledge, R.G.
    1987   Environmental cognition. In D. Stokols and I. Altman, eds., *Handbook of Environmental Psychology*. 2 volumes. New York: Wiley.

Gordon, H.
    1954   The economic theory of a common property resource: The fishery. *Journal of Political Economy* 62:124–142.

Gould, L.C., G.T. Gardner, D.R. DeLuca, A.R. Tiemann, L.W. Doob, and J.A.J. Stolwijk
    1988   *Perceptions of Technological Risks and Benefits*. New York: Russell Sage Foundation.

Gray, P.
    1989   The paradox of technological development. Pp. 192–204 in J.H. Ausubel and H.E. Sladovich, eds., *Technology and Environment*. Washington, D.C.: National Academy Press.

Greenberger, M.
    1983   *Caught Unawares: The Energy Decade in Retrospect*. New York: Harper Business.

Greenberger, M., M.A. Crenson, and B.L. Crissey
    1976   *Models in the Policy Process: Public Decision Making in the Computer Age*. New York: Russell Sage.

Groennings, S., E.W. Kelley, and M. Leiserson
    1970   *The Study of Coalition Behavior*. New York: Holt, Rinehart, and Winston.

Haas, E.B.
    1990   *When Knowledge is Power. Three Models of Change in*

International Organizations. Berkeley, Calif.: University of California Press.

Haas, P.M.
  1989   The regulation of CFCs and the Human Dimensions of Global Change. Unpublished paper prepared for the National Academy of Sciences' Committee on the Human Dimensions of Global Change, December.

Haas, P.M., ed.
  1991   Special issue on epistemic communities. *International Organization*, in press.

Habermas, J.
  1970   *Toward a Rational Society.* Boston, Mass.: Beacon.
  1981   New social movements. *Telos* 49:33–37.

Haefele, W., H. Barnert, S. Messner, M. Strubegger, and J. Anderer
  1986   Novel integrated energy systems: The case of zero emissions. Pp. 171–193 in W.C. Clark and R.E. Munn, eds., *Sustainable Development of the Biosphere.* New York: Cambridge University Press.

Haigh, N.
  1989   New tools for European air pollution control. *International Environmental Affairs* 1(1):26–38.

Hall, H.
  1888   The ice industry in the United States, with a brief sketch of its history and estimates of production in the United States. In U.S. Department of the Interior, Census Office, Tenth Census of the United States, 1880 (Washington, 1888), 22:1–5.

Hammond, K.R.
  1965   New directions in research on conflict resolution. *Journal of Social Issues* 11:44–66.

Hammond, K.R., T.R. Stewart, L. Adelman, and N. Wascoe
  1975   Report to the Denver City Council and Mayor regarding the choice of handgun ammunition for the Denver Police Department. Boulder, Colo.: University of Colorado, Program on Judgment and Social Interaction, Report No. 179.

Hansen, M.H., W.N. Hurwitz, and W. Madow
  1953   *Sample Survey Methods and Theory.* Volumes 1 and 2. New York: Wiley.

Hardin, G.
  1968   The tragedy of the commons. *Science* 162:1243–1248.

Hardin, R.
  1982   *Collective Action.* Baltimore, Md.: Johns Hopkins.

Harvey, D.
  1974   Population, resources, and the ideology of science. *Economic Geography* 50:256–277.

Heathcoate, R.L.
  1985   Extreme event analysis. Pp. 369–401 in R.W. Kates, J.H. Ausubel, and M. Berberian, eds., *Climate Impact Assess-*

ment: *Studies of the Interaction of Climate and Society.* ICSU/SCOPE Report No. 27. Chichester: John Wiley.

Heathcoate, R.L., and J.A. Mabbutt, eds.
1988   *Land, Water, and People.* Sydney: Allen and Unwin.

Heberlein, T.A.
1977   Norm activation and environmental action. *Journal of Social Issues* 33(3):207–211.

Hecht, S.B.
1981   Deforestation in the Amazon basin: Magnitude, dynamics, and soil resource effects. *Studies in Third World Societies* 13:61–108.
1985   Environment, development, and politics: Capital accumulation and the livestock sector in eastern Amazonia. *World Development* 13(6):663–684.
1989a Indigenous soil management in the Amazon basin: Some implications for development. Pp. 166–181 in J.O. Browder, ed., *Fragile Lands in Latin America.* Boulder, Colo.: Westview.
1989b The sacred cow in the green hell: Livestock and forest conversion in the Brazilian Amazon. *The Ecologist* 19:229–234.

Hecht, S.B., and A. Cockburn
1989   *The Fate of the Forest.* London: Verso.

Herman, R., S.A. Ardekani, and J.H. Ausubel
1989   Dematerialization. Pp. 50–69 in J.H. Ausubel and H.E. Sladovich, eds., *Technology and Environment.* Washington, D.C.: National Academy Press.

Hewitt, K., ed.
1983   *Interpretations of Calamity from the Viewpoint of Human Ecology.* Boston, Mass.: Allen and Unwin.

Hildyard, N.
1990   Adios Amazonia? A Report from the Altimira gathering. *The Ecologist* 10:53–62.

Hohenemser, C., R.E. Kasperson, and R.W. Kates
1985   Causal structure. Pp. 25–42 in R.W. Kates, C. Hohenemser, and J.X. Kasperson, eds., *Perilous Progress: Managing the Hazards of Technology.* Boulder, Colo.: Westview.

Holdren, J., K. Anderson, P.H. Gleick, I. Mintzer, G. Morris, and K.R. Smith
1979   Risk of Renewable Energy Sources: A Critique of the Inhaber Report. Report No. ERG 79-3. Berkeley, Calif.: University of California, Energy Resources Group.

Holling, C.S.
1978   *Adaptive Environmental Assessment and Management.* New York: Wiley.
1986   The resilience of terrestrial ecosystems: Local surprise and global change. Pp. 292–317 in W.C. Clark and R.E. Mucc, eds. *Sustainable Development of the Biosphere.* New York: Cambridge University Press.

Holusha, J.
  1990a   Ozone issue: Economics of a ban. *New York Times* January 11:D1, D8.
  1990b   Dupont to construct plants for ozone safe refrigerant. *New York Times* June 23:31–32.
Homer-Dixon, T.F.
  1990   Environmental Change and Violent Conflict. Occasional Paper No. 4, International Security Studies Program, American Academy of Arts and Sciences, Cambridge, Mass.
Hoos, I.R.
  1972   *Systems Analysis in Public Policy.* Berkeley, Calif.: University of California Press.
Hopmann, P.T.
  1978   Asymmetrical bargaining in the conference on security and cooperation in Europe. *International Organization* 32:141–177.
Houghton, J.T., G.J. Jenkins, and J.J. Ephraums, eds.
  1990   *Climate Change: The IPCC Scientific Assessment.* New York: Cambridge University Press.
Inglehart, R.
  1990   *Culture Shift in Advanced Industrial Society.* Princeton, N.J.: Princeton University Press.
Inhaber, H.
  1978   *Risk of Energy Production,* 2nd ed. 1119 REV-1. Ottawa: Atomic Energy Control Board.
International Energy Agency
  1987   *Energy Conservation in IEA Countries.* Paris: Organization for Economic Co-operation and Development.
International Federation of Institutes for Advanced Study
  1987   *The Human Response to Global Change: Prospectus for an International Research Program.* Toronto: Author.
Jacobson, H., and M. Price
  1990   *A Framework for Research on the Human Dimensions of Global Environmental Change.* ISSC/UNESCO Series 3.
Jasanoff, S.
  1986   *Risk Management and Political Culture.* New York: Russell Sage.
  1990   *The Fifth Branch: Science Advisers as Policy Makers.* Cambridge, Mass.: Harvard University Press.
Jasper, J.M.
  1988   The political life cycle of technological controversy. *Social Forces* 67:357–377.
  1990   *Nuclear Politics: Energy and the State in the United States, Sweden, and France.* Princeton, N.J.: Princeton University Press.
Jodha, N.S., and A.S. Mascarenhas
  1985   Adjustment in self-provisioning societies. In R.W. Kates,

J.H. Ausubel, and M. Berberian, eds., *Climate Impact Assessment.* SCOPE 27. Chichester: John Wiley & Sons.

Kahneman, D., P. Slovic, and A. Tversky
1982 *Judgment Under Uncertainty: Heuristics and Biases.* New York: Cambridge University Press.

Kaplan, M.A.
1989 Steps towards a democratic world order. *International Journal of World Peace* 6:23–44.

Kasprzyk, L.
1989 Science and technology policy and global change. *International Social Science Journal* 41:433–439.

Kates, R.W.
1971 Natural hazards in human ecological perspective: Hypotheses and models. *Economic Geography* 47:438–451.
1981 Drought in the Sahel. *Mazingira* 5(2):72–83.
1985a The human use of the biosphere. Pp. 491–493 in T.F. Malone and J.G. Roederer, eds., *Global Change.* New York: Cambridge University Press.
1985b The interaction of climate and society. Pp. 3–36 in R.W. Kates, J.H. Ausubel, and M. Berberian, eds., *Climate Impact Assessment.* New York: Wiley.

Kates, R.W., J.H. Ausubel, and M. Berberian, eds.
1985 *Climate Impact Assessment: Studies of the Interaction of Climate and Society.* ICSU/SCOPE Report No. 27. Chichester: John Wiley.

Katzev, R.D., and T.R. Johnson
1987 *Promoting Energy Conservation: An Analysis of Behavioral Research.* Boulder, Colo.: Westview.

Keeney, R.L., and H. Raiffa
1976 *Decisions With Multiple Objectives: Preferences and Value Tradeoffs.* New York: Wiley.

Kelman, M.
1987 *A Guide to Critical Legal Studies.* Cambridge, Mass.: Harvard University Press.

Kempton, W.
1991 Lay perspectives on global climate change. *Global Environmental Change* 1:183–208.

Keohane, R.O.
1984 *After Hegemony: Cooperation and Discord in the World Political Economy.* Princeton, N.J.: Princeton University Press.

Kikkaw, J.
1986 Complexity, diversity, and stability. Pp. 41–62 in J. Kikkaw and D.J. Anderson, *Community Ecology.* Melbourne: Blackwell.

King, K., and M. Chandler
1978 *The Wasted Lands.* Nairobi: ICRAF.

Kingdon, J.W.
  1984   *Agendas, Alternatives, and Public Policies.* Glenview, Ill.:
          Scott, Foresman.
Kinzelbach, W.K.H.
  1989   Energy and environment in China. *Environmental Policy
          and Law* 8:78–82.
Kish, L.
  1965   *Survey Sampling.* New York: Wiley.
Kmenta, J., and J.B. Ramsey, eds.
  1982   *Evaluating the Reliability of Macro-Economic Models.* New
          York: Wiley.
Kneese, A., and B. Bower
  1979   *Environmental Quality and Residuals Management.* Balti-
          more, Md.: Johns Hopkins University Press.
Kneese, A.V, and C.S. Russell
  1987   Environmental economics. Pp. 159–164 in J. Eatwell, M.
          Milgate, and P. Newman, eds., *The New Palgrave: A Dic-
          tionary of Economics*, Volume 2. London: Macmillan.
Kotlyakov, V.M., J.R. Mather, G.V. Sdasyuk, and G.F. White
  1988   Global change: Geographic approaches (A review). *Pro-
          ceedings of the National Academy of Sciences* (US) 85:5986–
          5991.
Krasner, S., ed.
  1983   *International Regimes.* Ithaca, N.Y.: Cornell University Press.
  1989   Sovereignty: An institutional perspective. Pp. 69–96 in
          J.A. Caporaso, ed., *The Elusive State: International and
          Comparative Perspectives.* Newbury Park, Calif.: Sage.
Kroll-Smith, J.S., S.R. Crouch, and A.G. Levine
  1991   Technological hazards and disasters. In R.E. Dunlap and
          W. Michelson, eds., *Handbook of Environmental Sociology.*
          Greenwich, Conn.: Greenwood.
Krutilla, J., and A. Fisher
  1975   *The Economics of Natural Environments.* Washington, D.C.:
          Resources for the Future.
Kujovich, M.Y.
  1970   The refrigerator car and the growth of the American dressed
          beef industry. *Business History Review* 44:460–482.
Kuznets, S.
  1983   *Economic Change: Selected Essays in Business Cycles, Na-
          tional Income, and Economic Growth*, reprint of 1953 edi-
          tion. Westport, Conn.: Greenwood.
Lal, R., P.A. Sanchez, and R.W. Cummings, Jr., eds.
  1986   *Land Clearing and Development in the Tropics.* Rotterdam,
          The Netherlands: A.A. Balkema.
Land, K.C., and S.H. Schneider
  1987   *Forecasting in the Social and Natural Sciences.* Dordrecht,
          The Netherlands: Reidel.

Lawrence, L.E.
  1965   The Wisconsin ice trade. *Wisconsin Magazine of History* 48:257–267.

Lee, R.D.
  1978   *Econometric Studies of Topics in Demographic History.* New York: Arno Press.

Leistritz, F.L., and S.H. Murdock
  1981   *The Socioeconomic Impact of Resource Development: Methods for Assessment.* Boulder, Colo.: Westview.

Leone, R.A.
  1987   *Who Profits: Winners, Losers, and Government Regulation.* New York: Basic Books.

Levins, R.
  1966   The strategy of model building in population biology. *American Scientist* 54:421–431.

Lichtenstein, S., P. Slovic, B. Fischhoff, M. Layman, and B. Combs
  1978   Judged frequency of lethal events. *Journal of Experimental Psychology: Human Learning and Memory* 4:551–578.

Linares, O.F.
  1976   "Garden hunting" in the American tropics. *Human Ecology* 4:331–349.

Lind, R.C., ed.
  1986   *Discounting for Time and Risk in Energy Policy.* Baltimore: Johns Hopkins University Press.

Lindblom, C.E., and D.K. Cohen
  1979   *Usable Knowledge.* New Haven, Conn.: Yale University Press.

Lindert, P.H.
  1978   *Fertility and Scarcity in America.* Princeton, N.J.: Princeton University Press.

Lipsey, R.G., and K. Lancaster
  1956   The general theory of second best. *Review of Economic Studies* 24:11–32.

Lipton, M.
  1977   *Why Poor People Stay Poor: Urban Bias in World Development.* Cambridge, Mass.: Harvard University Press.

Liverman, D.M., M.E. Hanson, B.J. Brown, and R.W. Merideth, Jr.
  1988   Global sustainability: Toward measurement. *Environmental Management* 12:133–143.

Lloyd, P.
  1991   Iron determinations. *Nature* 350:19.

Lyster, S.
  1985   *International Wildlife Law.* Cambridge: Grotius.

Macdonald, T.
  1981   Indigenous resource to an expanding frontier: Jungle Quichua economic conversion to cattle ranching. In N. Whitten, ed., *Cultural Transformations and Ethnicity in Modern Ecuador.* Urbana: University of Illinois Press.

Machlis, G.E., and J.E. Force
    1988   Community stability and timber-dependent communities. *Rural Sociology* 53:221–234.
Machlis, G.E., J.E. Force, and R.G. Balice
    1990   Timber, minerals, and social change: An exploratory test of two resource-dependent communities. *Rural Sociology* 55:411–424.
MacLean, D.E.
    1990   Comparing values in environmental policies: Moral issues and moral arguments. In P.B. Hammond and R. Coppock, eds., *Valuing Health Risks, Costs, and Benefits for Environmental Decision Making.* Washington, D.C.: National Academy Press.
Mahar, D.
    1988   *Government Policies and Deforestation in Brazil's Amazon Region.* Washington, D.C.: World Bank.
Majone, G.
    1989   *Evidence, Argument, and Persuasion in the Policy Process.* New Haven, Conn.: Yale University Press.
Mangun, W.
    1979   West European institutional arrangements for environmental policy implementation—especially air and water pollution control. *Environmental Conservation* 6(3):201–211.
March, J.G., and J.P. Olsen
    1989   *Rediscovering Institutions: The Organizational Basis of Politics.* New York: Free Press.
Marsh, G.P.
    1864   *Man and Nature; or, the Earth as Modified by Human Action.* New York: Scribner's.
Martin, J.H., R.M. Gordon, and S.E. Fitzwater
    1990   Iron in Antarctic waters. *Nature* 345:156–158.
Masini, E.
    1984   Futures research and global change. *Futures* 16:468–470.
Mather, J.R., and G.V. Sdasyuk
    1990   Global change: Some concepts and problems of geographical research. *GeoJournal* 20(2):8–9.
Mathews, J.T.
    1989   Redefining security. *Foreign Affairs* 68(2):162–177.
May, P.J.
    1985   *Recovery from Catastrophes.* Westport, Conn.: Greenwood.
May, R.
    1973   *Stability and Complexity in Model Ecosystems.* Princeton, N.J.: Princeton University Press.
Mazur, A.
    1981   *The Dynamics of Technical Controversy.* Washington, D.C.: Communications Press.
McEvoy, A.F.
    1986   *The Fisherman's Problem: Ecology and Law in the California Fisheries, 1850–1980.* New York: Cambridge University Press.

Mead, G.H.
1934   *Mind, Self, and Society.* Chicago, Ill.: University of Chicago Press.
Meadows, D., and J. Robinson
1985   *The Electronic Oracle: Computer Models and Social Decisions.* New York: Wiley.
Meadows, D.H.
1985   Charting the way the world works. *Technology Review* (Feb./Mar.):55.63.
Medvedev, Z.A.
1990   The environmental destruction of the Soviet Union. *The Ecologist* 20:24–29.
Meidinger, E., and A. Schnaiberg
1980   Social impact assessment as evaluation research: Claimants and claims. *Evaluation Review* 4:507–536.
Meiners, R.E., and B. Yandle
1989   *Regulation and the Reagan Era: Politics, Bureaucracy, and the Public Interest.* New York: Holmes and Meier.
Merchant, C.
1980   *The Death of Nature: Women, Ecology, and the Scientific Revolution.* New York: Harper and Row.
1991   The realm of social relations: Production, reproduction, and gender in environmental transformations. Pp. 673–684 in B.L. Turner et al., eds., *The Earth as Transformed by Human Action.* New York: Cambridge University Press.
Merton, R.K.
1949   *Social Theory and Social Structure.* Glencoe, Ill.: Free Press.
Miles, I., H. Rush, K. Turner, and J. Bessant
1988   *Information Horizons: The Long-term Social Implications of New Information Technology.* Brookfield, Vt.: Gower Publishing Company.
Mileti, D.S., and C. Fitzpatrick
1991   Communication of public risk: Its theory and its application. *Sociological Practice Review* 2(1):20–28.
Mileti, D.S., and J.M. Nigg
1991   Natural hazards and disasters. In R.E. Dunlap and W. Michelson, eds., *Handbook of Environmental Sociology.* Greenwich, Conn.: Greenwood.
Mishan, E.J.
1971   The postwar literature on externalities: an interpretive essay. *Journal of Economic Literature* 9:1–28
Mitchell, J.K.
1989   Hazards research. In G.L. Gaile and C.J. Wilmott, eds., *Geography in America.* Columbus, Ohio: Merrill.
Mitchell, R.C., and R.T. Carson
1988   *Using Surveys to Value Public Goods: The Contingent Valuation Method.* Washington, D.C.. Resources for the Future.

Modelski, G.
  1987   *Exploring Long Cycles.* Boulder, Colo.: Lynne Rienner Publishers.
Molina, M.J., and F.S. Rowland
  1974   Strato sink-chlorofluoromethane-chlorine atom ozone kill. *Nature* 249:810–814.
Moran, E.
  1976   Agricultural Development in the Transamazon Highway. Latin American Studies Working Papers. Bloomington: Indiana University.
  1987   Monitoring fertility degradation of agricultural lands in the lowland tropics. Pp. 69–91 in P.D. Little, M.M. Horowitz, A.E. Nyerges, eds., *Lands at Risk in the Third World: Local Level Perspectives.* Boulder, Colo.: Westview.
  1989   Amazon Soils: Distribution and Agricultural Alternatives under Indigenous and Contemporary Management. Paper for Wenner-Gren Symposium "Amazonian Synthesis."
Morrisette, P.M.
  1988   The stability bias and adjustment to climatic variability: The case of the rising level of the Great Salt Lake. *Applied Geography* 8:171–189.
  1989   The evolution of policy responses to stratospheric ozone depletion. *Natural Resources Journal* 29(3 Summer):793–820.
Morrisette, P.M., J. Darmstadter, A.J. Plantinga, and M.A. Toman
  1990   Lessons From Other International Agreements for a Global CO_2 Accord. Discussion paper ENR91-02, October. Washington, D.C.: Resources for the Future.
Morrison, D.E.
  1991   The environmental movement. In R.E. Dunlap and W. Michelson, eds., *Handbook of Environmental Sociology.* Greenwich, Conn.: Greenwood.
Mortimore, M.
  1989   Drought and drought response in the Sahel. Background paper prepared for the Committee on Human Dimensions of Global Change, National Research Council.
Mowlana, H., and L.J. Wilson
  1990   The passing of modernity: Communication and the transformation of society. Unpublished paper, University of Pennsylvania.
Muller, E.N.
  1988   Democracy, economic development, and income inequality. *American Sociological Review* 53:50–68.
Nash, M.
  1989   *The Cauldron of Ethnicity in the Modern World.* Chicago: University of Chicago Press.

National Academy of Sciences
 1991a *Policy Implications of Greenhouse Warming.* Washington, D.C.: National Academy Press.
 1991b *Policy Implications of Greenhouse Warming: Report of the Mitigation Panel.* Washington, D.C.: National Academy Press.

National Research Council
 1985 *Reducing Hazardous Waste Generation: An Evaluation and a Call for Action.* Washington, D.C.: National Academy Press.
 1986 *Population Growth and Economic Development: Policy Questions.* Washington, D.C.: National Academy Press.
 1987 *Confronting Natural Disasters.* Washington, D.C.: National Academy Press.
 1989a *Alternative Agriculture.* Washington, D.C.: National Academy Press.
 1989b *Improving Risk Communication.* Committee on Risk Perception and Communication, Washington, D.C.: National Academy Press.
 1990a *Confronting Climate Change: Strategies for Energy Research and Development.* Energy Engineering Board. Washington, D.C.: National Academy Press.
 1990b *The U.S. Global Change Research Program: An Assessment of the FY 1991 Plans.* Washington, D.C.: National Academy Press.
 1990c *Research Strategies for the U.S. Global Change Research Program.* Washington, D.C.: National Academy Press.

Nelkin, D.
 1979 *Controversy: Politics of Technical Decisions.* Beverly Hills, Calif.: Sage.
 1988 *Selling Science: How the Press Covers Science and Technology.* New York: Freeman.

Netting, R.McC.
 1968 *Hill Farmers of Nigeria: Cultural Ecology of the Kofyar of the Jos Plateau.* Seattle: University of Washington Press.

Neyhart, L.A.
 1952 *Giant of the Yards.* Boston, Mass.: Houghton Mifflin.

Nordhaus, W.
 1990 To Curb or Not to Curb: The Economics of the Greenhouse Effect. Paper presented to the Annual Meeting of the American Association for the Advancement of Science, New Orleans, February.

Norse, E.A., K.L. Rosenbaum, D.S. Wilcove, B.A. Wilcox, W.H. Romme, D.W. Johnston, and M.L. Stout
 1986 *Conserving Biological Diversity in Our National Forests.* Washington, D.C.: Wilderness Society.

North, D.C.
 1981 *Structure and Change in Economic History.* New York: Norton.

North, D.C., and R.P. Thomas
    1973   *The Rise of the Western World: A New Economic History.*
            New York: Cambridge University Press.
Nuclear Regulatory Commission
    1978   Risk Assessment Review Group to the U.S. Nuclear Regula-
            tory Commission. NUREG/CR-0400. Washington, D.C.:
            U.S. Nuclear Regulatory Commission.
Nye, D.H., and D.J. Greenland
    1966   The Soil Under Shifting Cultivation. Technical Communi-
            cation No. 51, Commonwealth Bureau of Soils. Harpenden,
            U.K.: Commonwealth Agricultural Bureau.
Offe, C.
    1985   New social movements: Challenging the boundaries of in-
            stitutional politics. *Social Research* 52:817–868.
Olson, M.
    1965   *The Logic of Collective Action.* Cambridge, Mass.: Harvard
            University Press.
Oye, K.A.
    1986   Enhancing cooperation under anarchy: Hypotheses and strat-
            egies. Pp. 1–24 in K.A. Oye, ed., *Cooperation Under Anar-
            chy.* Princeton, N.J.: Princeton University Press.
Ozawa, C.P., and L. Susskind
    1985   Mediating science-intensive policy disputes. *Journal of Policy
            Analysis and Management* 5:23–39.
Parkin, S.
    1989   *Green Parties: An International Guide.* London: Heretic Books.
Parry, M.L., T.R. Carter, and N.T. Konijn, eds.
    1988   *The Impact of Climatic Variation on Agriculture,* 2 vol-
            umes. Dordrecht, The Netherlands: Kluwer.
Partridge, W.L.
    1984   The humid tropics cattle ranching complex: Cases from
            Panama reviewed. *Human Organization* 43:76–80.
Pearce, D.W., and R.K. Turner
    1990   *Economics of Natural Resources and the Environment.* Bal-
            timore, Md.: Johns Hopkins University Press.
Pimm, S.C.
    1982   *Food Webs.* London: Chapman and Hill.
Plotkin, M.J.
    1988   The outlook for new agricultural and industrial products
            from the tropics. Pp. 106–116 in E.O. Wilson, ed., *Biodiversity.*
            Washington, D.C.: National Academy Press.
Polanyi, K.
    1944   *The Great Transformation: The Political and Economic
            Origins of Our Time.* Boston, Mass.: Beacon.
Poole, P.
    1989   Developing a Partnership of Indigenous Peoples, Conserva-
            tionists, and Land Use Planners in Latin America. Policy,

Planning, and Research Working Paper-Environment. Washington, D.C.: World Bank.

Posey, D.A.
1983 Indigenous ecological knowledge and development of the Amazon. Pp. 225–257 in E. Moran, ed., *The Dilemma of Amazon Development*. Boulder, Colo.: Westview.
1989 Alternatives to forest destruction: Lessons from the Mebengokre Indians. *The Ecologist* 19:241–244.

Prance, G.
1979 Notes on the vegetation of Amazonia Brazil. Part 3: The terminology of Amazonian forest types subject to inundation. *Brittonia* 31(1):26–38.
1989 Economic prospects from tropical rainforest ethnobotany. Pp. 61–74 in J.O. Browder, ed., *Fragile Lands of Latin America*. Boulder, Colo.: Westview.

Pressman, J.L., and A. Wildavsky
1984 *Implementation*, 3rd ed. Berkeley: University of California Press.

Price, D.
1989 *Before the Bulldozer: The Nambiquara Indians and the World Bank*. Cabin John, Md.: Seven Locks.

Pruitt, D.G., and M.J. Kimmel
1977 Twenty years of experimental gaming: Critique, synthesis, and suggestions for the future. *Annual Review of Psychology* 28:363–392.

Puccia, C.J., and R. Levins
1985 *Qualitative Modeling of Complex Systems: An Introduction to Loop Analysis and Time Averaging*. Cambridge, Mass.: Harvard University Press.

Putnam, R.D.
1988 Diplomacy and domestic politics: The logic of two-level games. *International Organization* 42:427–460.

Pyne, S.
1982 *Fire in America*. Princeton, N.J.: Princeton University Press.

Quarantelli, E., and R.R. Dynes
1977 Response to social crises and disaster. *Annual Review of Sociology* 3:23–49.

Rabb, T.K.
1983 Climate and society in history: A research agenda (with bibliography). Pp. 62–115 in R.S. Chen, E. Boulding, and S.H. Schneider, eds., *Social Science Research and Climate Change: An Interdisciplinary Appraisal*. Dordrecht, The Netherlands: Reidel.

Rapoport, A.
1960 *Fights, Games, and Debates*. Ann Arbor, Mich.: University of Michigan Press.
1964 *Strategy and Conscience*. New York: Harper & Row.

Rappaport, R.A.
  1967   *Pigs for the Ancestors: Ritual in the Ecology of a New Guinea People.* New Haven, Conn.: Yale University Press.
Redclift, M.
  1987   *Sustainable Development: Exploring the Contradictions.* London: Methuen.
Rees, J.
  1985   *Natural Resources: Allocation, Economics and Policy.* London: Methuen.
Repetto, R., W. Magrath, M. Wells, C. Beer, and F. Rossini
  1989   *Wasting Assets: Natural Resources in the National Income Accounts.* Washington, D.C.: World Resources Institute.
Richards, J.F.
  1986   World environmental history and economic development. Pp. 53–71 in W.C. Clark and R.E. Munn, eds., *Sustainable Development of the Biosphere.* New York: Cambridge University Press.
Ridker, R.G.
  1972a Population and pollution in the United States. *Science* 176:1085–1090.
  1972b *Population, Resources and the Environment.* Washington, D.C.: U.S. Government Printing Office.
Riebsame, W.E.
  1987   Human response to climate change: The role of decision-maker perception. In *Proceedings of the Symposium on Climate Change in the Southern United States: Future Impacts and Present Policy Issues.*
Riebsame, W.E., H.F. Diaz, T. Moses, and M. Price
  1986   The social burden of weather and climate hazards. *Bulletin of the American Meteorological Society* 67(11):1378–1388.
Roan, S.
  1989   *Ozone Crisis: The Fifteen Year Evolution of a Sudden Global Emergency.* New York: Wiley.
Roberts, B.
  1990   Human rights and national security. *Washington Q* (Spring) (13):65–75.
Robinson, J.B.
  1982   Apples and horned toads: On the framework-determined nature of the energy debate. *Policy Sciences* 15:23–45.
  1988   Unlearning and backcasting: Rethinking some of the questions we ask about the future. *Technological Forecasting and Social Change* 33:325–338.
  1989   The Proof of the Pudding: Policy and Implementation Issues Associated with Increasing Energy Efficiency. Department of Environment and Resource Studies, University of Waterloo, November.

Rogers, E.M.
   1983   *Diffusion of Innovations.* New York: Free Press.
Rogers, E., and D.L. Kincaid
   1981   *Communication Networks: Toward a New Paradigm for Research.* New York: Free Press.
Rohrschneider, R.
   1990   The roots of public opinion toward new social movements: An empirical test of competing explanations. *American Journal of Political Science* 34:1–30.
Rosa, E.A., G.E. Machlis, and K.M. Keating
   1988   Energy and society. *Annual Review of Sociology* 14:149–172.
Rosenbaum, W.
   1991   *Environmental Politics and Policy,* 2nd ed. Washington, D.C.: CQ Press.
Rosenberg, N., P. Crosson, W. Easterling, M. McKenney, and K. Frederick
   1990   *Methodology for Assessing Regional Economic Impacts and Responses to Climate Change - The MINK Study.* Washington, D.C.: Resources for the Future.
Ross, L.
   1987   Environmental policy in post-Mao China. *Environment* 29(4):12–17, 34–39.
Rosswall, T.R., R.G. Woodmansee, and P.G. Risser
   1988   *SCOPE 35: Scales and Global Change: Spatial and Temporal Variability in Biospheric and Geospheric Processes.* Chichester: John Wiley & Sons.
Rostow, W.W.
   1960   *The Stages of Economic Growth: A Non-Communist Manifesto.* New York: Cambridge University Press.
   1978   *The World Economy: History and Prospect.* Austin, Tex.: University of Texas Press.
Roszak, T.
   1972   *Where the Wasteland Ends.* New York: Doubleday.
Rudel, T.K.
   1989   Population, development, and tropical deforestation: A cross-national study. *Rural Sociology* 54:327–338.
Russell, D.
   1987   Rush to market; biotechnology and agriculture. *The Amicus Journal* 9(1):16–38.
Ruttan, V.W.
   1971   Technology and the environment. *American Journal of Agricultural Economics* 53:707–717.
Ryder, N.B.
   1965   The cohort as a concept in the study of social change. *American Sociological Review* 30:843–861.
Saarinen, T.F.
   1982   *Perspectives on Increasing Hazard Awareness.* Monograph

35. Natural Hazards Research and Applications Information Center. Boulder, Colo.: University of Colorado.

Sack, R.D.
1990   The realm of meaning: The inadequacy of human nature theory and the view of mass consumption. Pp. 659–671 in B.L. Turner II et al., eds., *The Earth as Transformed by Human Action*. New York: Cambridge University Press.

Sagoff, M.
1988   *The Economy of the Earth*. New York: Cambridge University Press.

Sahal, D.
1981   *Patterns of Technological Innovation*. Reading, Mass.: Addison-Wesley Publishing Company.

Salati, E., and P.B. Vose
1984   Amazon basin: A system in equilibrium. *Science* 225:129–138.

Sanchez, P.A., D.E. Bandy, J.H. Villachica, and J.J. Nicholaides
1982   Amazon basin soils: Management for continuous crop production. *Science* 216:821–827.

Sand, P.H.
1990a   International cooperation: The environmental experience. In J.T. Mathews, ed., *Preserving the Global Environment: The Challenge of Shared Leadership*. New York: Norton.
1990b   *Lessons Learned in Global Environmental Governance*. Washington, D.C.: World Resources Institute.

Sandman, P.M., D.B. Sachsman, M.R. Greenberg, and M. Gochfeld
1987   *Environmental Risk and the Press*. New Brunswick, N.J.: Transaction Books.

Saunders, H.H.
1989   Changing Relationships: Beyond Negotiation in Resolving Conflict. Unpublished paper, The Brookings Institution.

Savory, A.
1988   *Holistic Resource Management*. Covelo, Calif.: Island Press.

Schipper, L.
1989   International Comparisons of Energy Efficiency. Unpublished memorandum, Lawrence Berkeley Laboratory, Berkeley, Calif., November 30.

Schipper, L., S. Bartlett, D. Hawk, and E. Vine
1989   Linking life-styles and energy use: A matter of time? *Annual Review of Energy* 14:273–320.

Schipper, L., R. Howarth, and H. Geller
1990   United States energy use from 1973 to 1987. *Annual Review of Energy* 15:455–504.

Schipper, L., A. Ketoff, and A. Kahane
1985   Explaining residential energy use using international bottom up comparisons. *Annual Review of Energy* 10:341–405.

Schmink, M., and C.H. Wood
    1987   The "political ecology" of Amazonia. Pp. 38–57 in P.D. Little, M.M. Horowitz, A.E. Nyerges, eds., *Lands at Risk in the Third World*. Boulder, Colo.: Westview.

Schnaiberg, A.
    1980   *The Environment: From Surplus to Scarcity*. New York: Oxford.
    1991   The political economy of environmental problems and policies: Consciousness, conflict, and control capacities. In R.E. Dunlap and W. Michelson, eds., *Handbook of Environmental Sociology*. Greenwich, Conn.: Greenwood.

Schneider, R.
    1990   *Environment and Development in the Amazon: Preliminary Observations and Back-to-Office Report*. Washington, D.C.: World Bank.

Schneider, S.H.
    1988   The whole earth dialogue. *Issues in Science and Technology* 4(3):93–99.

Schurr, S.H.
    1984   Energy use, technological change, and productive efficiency: An economic-historical interpretation. *Annual Review of Energy* 9:409–425.

Seidman, H., and R. Gilmour
    1986   *Politics, Position, and Power*. New York: Oxford.

Seligman, C.
    1989   Environmental ethics. *Journal of Social Issues* 45:169–184.

Sen, A.
    1982   Approaches for the choice of discount rates for social benefit-cost analyses. Pp. 325–353 in R.C. Lind, ed., *Discounting for Time and Risk in Energy Policy*. Washington, D.C.: Resources for the Future.

Sen, A.K.
    1981   *Poverty and Famines*. Oxford: Clarendon Press.

Shine, K. P., R.G. Derwent, D.J. Wuebbles, and J-J. Morcrette
    1990   Radiative forcing of climate. Pp. 41–68 in J.T. Houghton, G.J. Jenkins, and J.J. Ephraums, eds., *Climate Change: The IPCC Assessment*. New York: Cambridge University Press.

Shiva, V.
    1989   *Staying Alive: Women, Ecology, and Development*. London: Zed Books.

Simon, J.
    1981   *The Ultimate Resource*. Princeton, N.J.: Princeton University Press.

Simon, J.L., and H. Kahn, eds.
    1984   *The Resourceful Earth*. Oxford: Basil Blackwell.

Slovic, P., B. Fischhoff, S. Lichtenstein, B. Corrigan, and B. Combs

1977   Preference for insuring against probable small losses: Implications for the theory and practice of insurance. *Journal of Risk and Insurance* 44:237–258.

Slovic, P., H. Kunreuther, and G.F. White
1974   Decision processes, rationality, and adjustment to natural hazards. Pp. 187–205 in G.F. White, ed., *Natural Hazards: Local, National, Global.* New York: Oxford.

Slovic, P., B. Fischhoff, and S. Lichtenstein
1979   Rating the risks. *Environment* 21(3):14–20, 36–39.

Smil, V.
1988   *Energy in China's Modernization: Advances and Limitations.* Armonk, N.Y.: M.E. Sharpe.

Smith, T., and K. DeJong
1981   Genetic algorithms applied to the calibration of information driven models of U.S. migration patterns. Pp. 955–959 in *Proceedings of the 12th Annual Pittsburgh Conference on Modeling and Simulation.*

Smith, D.A., and B. London
1990   Convergence in world urbanization? A quantitative assessment. *Urban Affairs Quarterly* 25:574–590.

Smith, N.J.H.
1982   *Rainforest Corridors: The Transamazon Colonization Scheme.* Berkeley, Calif.: University of California Press.

Smith, V.K.
1979   *Scarcity and Growth Reconsidered.* Baltimore, Md.: Johns Hopkins Press for Resources of the Future.

Solomon, S.
1990   Progress towards a quantitative understanding of Antarctic ozone depletion. *Nature* 347:347–354.

Steenbergen, B.
1983   The sociologist as social architect. *Futures* 15:376–386.

Stein, J.G.
1989   *Getting to the Table.* Baltimore, Md.: Johns Hopkins University Press.

Stephens, E.H.
1989   Capitalist development and democracy in South America. *Political Sociology* 17:281–352.

Stern, P.C.
1986   Blind spots in policy analysis: What economics doesn't say about energy use. *Journal of Policy Analysis and Management* 5:200–227.
1991   Learning through conflict: A realistic strategy for risk communication. *Policy Sciences,* 24:99–119.

Stern, P.C., ed.
1984   *Improving Energy Demand Analysis.* Report of the Energy Demand Analysis Panel, Committee on Behavioral and Social

Aspects of Energy Consumption and Production, National Research Council. Washington, D.C.: National Academy Press.

Stern, P.C., and E. Aronson, eds.
1984 *Energy Use: The Human Dimension*. Report of the National Research Council Committee on the Behavioral and Social Aspects of Energy Consumption and Production. New York: Freeman.

Stern, P.C., E. Aronson, J.M. Darley, D.H. Hill, E. Hirst, W. Kempton, and T.J. Wilbanks
1986a The effectiveness of incentives for residential energy conservation. *Evaluation Review* 10:147–176.
1987 Answering behavioral questions about energy efficiency in buildings. *Energy* 12:339–353.

Stern, P.C., T. Dietz, and J.S. Black
1986b Support for environmental protection: The role of moral norms. *Population and Environment* 8:204–222.

Stern, P.C., and G.T. Gardner
1981a The place of behavior change in the management of environmental problems. *Zeitschrift für Umweltpolitik* 2:213–239.
1981b Psychological research and energy policy. *American Psychologist* 36:329–342.

Stern, P.C., and S. Oskamp
1987 Managing scarce environmental resources. Pp. 1043–1088 in D. Stokols and I. Altman, eds., *Handbook of Environmental Psychology*, Volume 2. New York: Wiley.

Steward, J.
1955 *Theory of Culture Change*. Urbana: University of Illinois Press.
1977 *Evolution and Ecology: Essays on Social Transformation*. Urbana: University of Illinois Press.

Stolarski, R.
1988 The Antarctic ozone hole. *Scientific American* (January)258:1.

Stone, C.D.
1987 *Earth and Other Ethics*. New York: Harper and Row.

Strauss, A.
1978 *Negotiations: Varieties, Contexts, Processes, and Social Order*. San Francisco: Jossey-Bass.

Subler, S., and C. Uhl
1990 Japanese agroforestry in Amazonia: A case study in Tomé-Açu, Brazil. Pp. 152–166 in A.B. Anderson, ed., *Sustainable Use of the Amazon Rain Forest*. New York: Columbia University Press.

Swift, L.F., and A. Van Vlissingen, Jr.
1927 *The Yankee of the Yards: The Biography of Gustavus Franklin Swift*. Chicago, Ill.: A.W. Shaw.

Syme, G.J., and E. Eaton
    1989   Public involvement as a negotiation process. *Journal of Social Issues* 45:87–107.
Tarlock, A.D., and P. Tarak
    1983   An overview of comparative environmental law. *Denver Journal of International Law and Policy* 13(1 Fall):85–108.
Thomas, W., Jr., ed.
    1956   *Man's Role in Changing the Face of the Earth.* Chicago, Ill.: University of Chicago Press.
Tietenberg, T.H.
    1985   *Emissions Trading: An Exercise in Reforming Pollution Policy.* Washington, D.C.: Resources for the Future.
Tilly, C.
    1989   *Big Structures, Large Processes, Huge Comparisons.* New York: Russell Sage.
    1990   *Coercion, Capital, and European States.* London: Basil Blackwell.
Timmerman, P.
    1986   Mythology and surprise in the sustainable development of the biosphere. Pp. 435–453 in W.C. Clark and R.E. Munn, eds., *Sustainable Development of the Biosphere.* New York: Cambridge University Press.
Toth, F.L.
    1988a   Policy exercises: Objectives and design elements. *Simulation and Games* 19:235–255.
    1988b   Policy exercises: Procedures and implementation. *Simulation and Games* 19:256–276.
Touraine, A.
    1985   An introduction to the study of social movements. *Social Research* 52:749–787.
Touraine, A., Z. Hegedus, F. Duket, and M. Wieviorka
    1983   *Anti-Nuclear Protest.* New York: Cambridge University Press.
Treece, D.
    1989   The militarization and industrialization of Amazonia. *The Ecologist* 19:225–228.
Tuan, Y.-F.
    1968   Discrepancies between environmental attitude and behavior: Examples from Europe and China. *The Canadian Geographer* 12:176–191.
Turner, B.L., II
    1989   The Human Causes of Global Environmental Change. In R.S. DeFries and T.F. Malone, eds., Global Change and Our Common Future: Papers from a Forum. National Academy of Sciences.
Turner, B.L., II, and S.B. Brush, eds.
    1987   *Comparative Farming Systems.* New York: Guilford Press.

Turner, B.L., II, and W.B. Meyer
    1991   Land use and land cover in global environmental change: Considerations for study. *International Social Science Journal*, in press.
Turner, B.L., II, G. Hyden, and R.W. Rates
    n.d.   Population Growth and Agricultural Intensification: Studies from Densely Settled Areas of Sub-Sagaran Africa. Book manuscript.
Turner, B.L., II, W.C. Clark, R.W. Kates, J.F. Richards, J.T. Mathews, and W.B. Meyer, eds.
    1991a   *The Earth as Transformed by Human Action*. New York: Cambridge University Press.
Turner, B.L., II, R.E. Kasperson, W.B. Meyer, K. Dow, D. Golding, J.X. Kasperson, R.C. Mitchell and S.J. Ratick
    1991b   Two types of global environmental change: Definitional and spatial scale issues in their human dimensions. *Global Environmental Change* 1(1):14–22.
Uhl, C., D. Nepstad, R. Buschbacher, K. Clark, B. Kauffman, and S. Subler
    1989   Disturbance and regeneration in Amazonia: Lessons for sustainable land use. *The Ecologist* 19:235–240.
Unfer, L.
    1951   Swift and Company: The Development of the Packing Industry, 1875 to 1912. Ph.D. Thesis, University of Illinois.
United Nations
    1989   Intergovernmental Committee on Science & Technology for Development. Innovations for development. *The Futurist* 23:48.
U.S. Department of Energy
    1989   *Energy Conservation Trends: Understanding the Factors that Affect Conservation Gains in the U.S. Economy*. DOE/PE-0092. Washington, D.C.: Author.
U.S. Office of Technology Assessment
    1986   *Serious Reduction of Hazardous Waste. For Pollution Prevention and Industrial Efficiency*. Washington, D.C.: U.S. Government Printing Office.
    1987   *Technologies to Maintain Biological Diversity*. Washington, D.C.: U.S. Government Printing Office.
    1988   *An Analysis of the Montreal Protocol on Substances that Deplete the Ozone Layer*. Washington, D.C.: U.S. Government Printing Office.
    1991   *Changing By Degrees: Steps to Reduce Greenhouse Gases*. Washington, D.C.: U.S. Government Printing Office.
Van Liere, K.D., and R.E. Dunlap
    1980   The social bases of environmental concern: A review of hypotheses, explanations, and empirical evidence. *Public Opinion Quarterly* 44:43–59, 181–197.
Vine, E., and J. Harris
    1988   *Planning for an Energy-Efficient Future. The Experience*

with *Implementing Energy Conservation Programs for New Residential and Commercial Buildings.* Volume 1. LBL-25525. Berkeley, Calif.: Lawrence Berkeley Laboratory.

Vine, E.L.
1981 *Solarizing America: The Davis Experience.* Washington, D.C.: Conference on Alternative State and Local Policies.

Vogel, D.
1986 *National Styles of Regulation.* Ithaca, N.Y.: Cornell University Press.
1990 Environmental policy in Europe and Japan. In N. Vig and M. Kraft, eds., *Environmental Policy in the 1990s.* Washington, D.C.: CQ Press.

von Winterfeldt, D. and W. Edwards
1984 Patterns of conflict about risky technologies. *Risk Analysis* 4(1):55–68.

Wallerstein, I.
1974 *The Modern World System: I.* New York: Academic Press.
1976 *The Modern World System: Capitalist Agriculture and the Origins of the European World Economy in the Sixteenth Century.* New York: Academic.
1980 *The Modern World System: II.* New York: Academic Press.
1988 *The Modern World System: III.* New York: Academic Press.

Watson, R.T., H. Rohde, H. Oeschger, and U. Siegenthaler
1990 Greenhouse gases and aerosols. Pp. 1–40 in J.T. Houghton, G.J. Jenkins, and J.J. Ephraums, eds., *Climate Change: The IPCC Assessment.* New York: Cambridge University Press.

Watts, M.
1987 *Silent Violence: Food, Famine, and Peasantry in Northern Nigeria.* Berkeley, Calif.: University of California Press.

Weber, M.
1958 *The Protestant Ethic and the Spirit of Capitalism,* trans. by Talcott Parsons. New York: Scribner's.

Weigel, R.H.
1977 Ideological and demographic correlates of proecology behavior. *Journal of Social Psychology* 103:39–47.

Weiss, C.H., and M.J. Bucuvalas
1980 *Social Science Research and Decision-Making.* New York: Columbia University Press.

Weiss, E.B.
1988 *In Fairness to Future Generations: International Law, Common Patrimony, and Intergenerational Equity.* New York: Transnational Publishers.

White, G.F., ed.
1974 *Natural Hazards: Local, National, Global.* New York: Oxford.

White, G.F., and J.E. Haas
1975 *Assessment of Research on Natural Hazards.* Cambridge, Mass.: MIT Press.

White, L.
    1967   The historical roots of our ecologic crisis. *Science* 155:1203–
            1207.
White, L.A.
    1959   *The Evolution of Culture: The Development of Civiliza-
            tion to the Fall of Rome.* New York: McGraw-Hill.
Whitmore, T., D. Johnson, B.L. Turner, II, R.W. Kates, and T. Gottschang
    1991   Long-term population change. Pp. 25–39 in B.L. Turner, II,
            et al., eds., *The Earth as Transformed by Human Action.*
            New York: Cambridge University Press.
Whyte, A.V.T.
    1985   Perception. Pp. 403–436 in R.W. Kates, J.H. Ausubel, and
            M. Berberian, eds., *Climate Impact Assessment: Studies of
            the Interaction of Climate and Society.* ICSU/SCOPE Re-
            port No. 27. Chichester: John Wiley.
    1986   From hazard perception to human ecology. Pp. 240–271 in
            R.W. Kates and I. Burton, eds., *Geography, Resources, and
            Environment, Vol. II.* Chicago, Ill.: University of Chicago
            Press.
Wilson, E.O.
    1988   The current state of biological diversity. Pp. 3–18 in E.O. Wil-
            son, ed., *Biodiversity.* Washington, D.C.: National Academy Press.
Wolf, E.R.
    1982   *Europe and the Peoples Without History.* Berkeley, Calif.:
            University of California Press.
World Bank
    1984   *World Development Report 1984.* New York: Oxford Uni-
            versity Press.
    1989   *World Development Report 1989.* New York: Oxford.
World Meteorological Organization
    1985   *Atmospheric Ozone 1985.* WMO Global Ozone Research
            and Monitoring Project Report No. 16.
World Resources Institute
    1985   *Tropical Forests: A Call to Action.* Washington, D.C.:
            World Resources Institute.
    1990   *World Resources, 1990–1991.* New York: Oxford.
World Resources Institute and International Institute for Environment
    and Development
    1988   *World Resources 1988–89.* New York: Basic.
Worster, D.
    1988   The vulnerable earth: Toward a planetary history. Pp. 3–
            20 in D. Worster, ed., *The Ends of the Earth: Perspectives
            on Modern Environmental History.* New York: Cambridge
            University Press.
    1989   The Dust Bowl of North America. Background paper pre-
            pared for the Committee on the Human Dimensions of
            Global Change, National Research Council.

Wright, J.R.
1990   Contributions, lobbying, and committee voting in the U.S. House of Representatives. *American Political Science Review* 84:417–438.

Wright, K.
1990   The road to the global village. *Scientific American* 262:83–85.

Wynne, B.
1984   The institutional context of science, models and policy. *Policy Sciences* 17:277–320.

Xi, X., E.S. Rubin, and M.G. Morgan
1989   Coal use in China and its environmental implications. Paper No. U.S. 23, Proceedings of Pittsburgh Coal Conference, September.

Yeager, M.
1981   *Competition and Regulation: The Development of the Oligopoly in the Meat Packing Industry.* Greenwich, Conn.: JAI Press.

Young, O.R.
1989a   *International Cooperation: Building Regimes for Natural Resources and the Environment.* Ithaca, N.Y.: Cornell University Press.
1989b   The politics of international regime formation: Managing natural resources and the environment. *International Organization* 43:349–375.

Zartman, I.W., and M.R. Berman
1982   *The Practical Negotiator.* New Haven, Conn.: Yale University Press.

Zinberg, D.S.
1983   *Uncertain Power: The Struggle for a National Energy Policy.* New York: Pergamon.

Index